AF817647

THE DYNAMICS OF MIGRATION, HEALTH AND LIVELIHOODS

Dr Kubaje Adazu
(1961–2009)

We would like to dedicate this book to the beloved memory of Dr. Kubaje Adazu, Chief Demographer and Technical Consultant of the Health Demographic Surveillance System (HDSS) of the Kenya Medical Research Institute/Centres for Disease Control and Prevention (CDC), Kisumu, Kenya and co-leader of the INDEPTH Migration and Urbanization Working Group. Dr. Adazu passed away at his home in Kisumu on 19 January 2009. He will be remembered and honoured for his great contribution to scientific knowledge and skills development in social demography. He was firmly convinced of the importance of understanding how migration influences families and this book is a testimony of the energy he put into migration research. At the time of his death Adazu was leading a team of researchers assessing the contribution of demographic factors to mortality and morbidity due to HIV/AIDS and other diseases of major public health importance. He was also active in the technical and editorial work required for the production of this book. We will always remember him for his guidance and skill and miss his friendship.

The Dynamics of Migration, Health and Livelihoods

INDEPTH Network Perspectives

Edited by

MARK COLLINSON

MRC/University of the Witwatersrand, RSA

KUBAJE ADAZU

Kenya Medical Research Institute, Kenya

MICHAEL WHITE

Brown University, USA

SALLY FINDLEY

Columbia University, USA

Routledge

Taylor & Francis Group

LONDON AND NEW YORK

First published 2009 by Ashgate Publishing

Reissued 2018 by Routledge
2 Park Square, Milton Park, Abingdon, Oxon OX14 4RN
711 Third Avenue, New York, NY 10017, USA

Routledge is an imprint of the Taylor & Francis Group, an informa business

Copyright © INDEPTH Network 2009

Mark Collinson, Kubaje Adazu, Michael White and Sally Findley have asserted their right under the Copyright, Designs and Patents Act, 1988, to be identified as the editors of this work.

All rights reserved. No part of this book may be reprinted or reproduced or utilised in any form or by any electronic, mechanical, or other means, now known or hereafter invented, including photocopying and recording, or in any information storage or retrieval system, without permission in writing from the publishers.

A Library of Congress record exists under LC control number: 2009030787

Notice:
Product or corporate names may be trademarks or registered trademarks, and are used only for identification and explanation without intent to infringe.

Publisher's Note
The publisher has gone to great lengths to ensure the quality of this reprint but points out that some imperfections in the original copies may be apparent.

Disclaimer
The publisher has made every effort to trace copyright holders and welcomes correspondence from those they have been unable to contact.

ISBN 13: 978-0-815-39767-0 (hbk)
ISBN 13: 978-1-351-14704-0 (ebk)

Contents

List of Figures — *vii*
List of Tables — *ix*
List of Abbreviations — *xi*
Notes on Contributors — *xiii*
Foreword — *xxv*
Preface — *xxvii*
Acknowledgements — *xxix*

PART I INTRODUCTION

1 Migration and Demographic Surveillance: An Overview of Opportunities and Challenges — 3
Michael J. White

2 Health and Demographic Surveillance Migration Methodology and Data: A Promise for Cross-Site Comparative Analyses — 19
Kubaje Adazu

3 Dynamics of Space and Time: Community Context of Migration, Livelihoods and Health in the INDEPTH Sites — 33
Sally E. Findley

4 Age–Sex Profiles of Migration: Who is a Migrant? — 49
Mark A. Collinson

PART II MIGRATION AND LIVELIHOODS

5 Migration and Agricultural Production in Kanchanaburi, Thailand — 65
Sureeporn Punpuing and Philip Guest

6 Migration and Socio-Economic Change in Rural South Africa, 2000–2007 — 81
Mark A. Collinson, Annette A.M. Gerritsen, Samuel J. Clark, Kathleen Kahn and Stephen M. Tollman

7 Parents' Migration and Children's Education in Matlab, Bangladesh 109
Nurul Alam and Peter Kim Streatfield

PART III MIGRATION AND HEALTH

8 Assessing the Effect of Mother's Migration on Childhood Mortality in the Informal Settlements of Nairobi 123
Adama Konseiga, Eliya M. Zulu, Philippe Bocquier, Kanyiva Muindi, Donatien Beguy and Yazoumé Yé

9 Child Migration and Mortality in Rural Nyanza Province: Evidence from the Kisumu Health and Demographic Surveillance System (KHDSS) in Western Kenya 139
Kubaje Adazu, Daniel Feiken, Peter Ofware, Bernard Onyango, David Obor, Rose Kiriinya, Laurence Slutsker, John Vulule and Kayla Laserson

10 Migration and Adult Mortality in Rural Southern Mozambique: Evidence from the Demographic Surveillance System in Manhiça District 159
Ariel Nhacolo, Delino Nhalungo, Charfudin Sacoor, Leonildo Matsinhe, John Aponte and Pedro Alonso

11 Migration and Under Five Morbidity in Bavi, Vietnam 169
Ho Dang Phuc, Nguyen Xuan Thanh and Nguyen Thi Kim Chuc

PART IV EPILOGUE

12 The Relevance of the Volume 185
Cheikh Mbacké

Index *193*

List of Figures

3.1 Integrated model of sustainable livelihoods, migration and health 37
4.1 Age–sex profiles of in- and out-migration from the Kisumu HDSS site in 2002 52
4.2 Age profiles of male out-migration from seven HDSS sites in 2002 54
4.3 Age profiles of male in-migration in seven HDSS sites in 2002 55
4.4 Age profiles of female out-migration from seven HDSS sites in 2002 56
4.5 Age profiles of female in-migration in seven HDSS sites in 2002 57
5.1 Mean number of consumer items owned by agricultural households by type of migrant household and year 71
5.2 Mean amount of land (in acres) used by agricultural households for intensive forms of agriculture, by type of migrant household and year 72
5.3 Mean amount of land (in acres) used by agricultural households for less intensive forms of agriculture, by type of migrant household and year 72
6.1 The distribution of socio-economic status. Histogram of the percentage of households by socio-economic status showing changes in poverty levels between three successive rounds of data collection from 2001–2005 90
9.1 Map of the study area – DSA, Nyanza Province, Western Kenya 144
10.1 Age-specific mortality rates by sex, Manhiça HDSS area 1998–2005 162
10.2 Age-specific mortality rates by migrant status, Manhiça HDSS area 1998–2005 162
10.3 39q20 from 1998–2005 by migrant status, Manhiça HDSS area 163
10.4 39q20 for migrants from 1998–2005 by sex, Manhiça HDSS area 163
11.1 Age and net migration by gender in Filabavi 174
11.2 Filabavi 2003 population pyramid 175

List of Figures

List of Tables

2.1	Site frequency of updates and definition of migration	24
3.1	Migration geography of INDEPTH sites	39
3.2	Environmental and agricultural context of the INDEPTH sites	40
3.3	Macro-economic macro-context of the INDEPTH sites	41
3.4	Community living standards at INDEPTH sites	44
3.5	Situational and social factors affecting migration timing	44
4.1	Migration rates by age and gender, by migration duration minimum threshold (one versus six months) (with p values for test of significant differences)	60
4.2	Migration rates by age and gender, by migration duration threshold (three versus four months) (p value for test of significant differences)	61
5.1	Percentage distribution of households by out-migration status and year (n=4,555)	71
5.2	Mean number of household members employed in agriculture by year and household migration status (n=4,955)	73
5.3	Unstandardized ordinary least squares regression coefficients of models predicting amount of land used in agriculture	74
5.4	Unstandardized ordinary least squares regression coefficients of models predicting proportion of household members working in agriculture	76
6.1	Type of remittance by sex and year, with p-values showing the sex differential in each year; and the trend 2002–2007 for each sex	92
6.2	Average monthly amount remitted by sex and year, with p-values showing the sex differential	93
6.3	Destinations of migrants by sex, with p-values showing the sex differential; and the percent of migrants of both sexes remitting from each destination category	94
6.4	OLS regression on household socio-economic status at the end of the period, by migration and other household factors from 2001–2005	98
6.5	Logistic regression on households that are not chronically poor from 2001–2005, by migration and other household factors	102
7.1	Distribution of children (n=63,208) aged 6–19 years by parents' migration destination during 1996–2004	111
7.2	Demographic and socioeconomic factors associated with children's school enrolment and passing grade five, Matlab HDSS area 2005	112

7.3	Demographic and socioeconomic factors associated with children's passing grade seven and grade ten, Matlab HDSS area 2005	114
7.4	Effects of parents' duration of out-migration (in months) on children's education, Matlab HDSS area 2005	115
7.5	Parents' migration status and percent of children aged 6–19 years who belong to top household asset quintiles (4th or 5th)	116
8.1	Basic social and demographic indicators from NUHDSS, 2003–2006	126
8.2	Attrition and mortality models for slum-born children under four (2003–2006)	130
9.1	Summary entry-age statistics, by entry type	153
9.2	Distribution of the children by selected background characteristics	153
9.3	Distribution of relative hazard of death by migration status and other selected covariates (excluding deaths in the first three months of life)	154
10.1	Factors associated with mortality among adult migrants, Manhiça 1998–2005 (multivariate analysis)	165
11.1	Migration by years in Filabavi	174
11.2	Reasons for migration by sex and type (numbers and percentages)	176
11.3	Frequency of migration by economic status, sex and type	176
11.4	Frequency of migration for economic reasons by economic status, sex and type	177
11.5	Reasons for migration by educational levels and type (numbers and percentages)	178
11.6	Frequency of migration for economic reasons by educational levels, sex and type	178
11.7	Reasons for migration by occupation and type (numbers and percentages)	179
11.8	Frequency of migration for economic reasons by occupation, sex and type	179
11.9	Determinants of under five morbidity	181

List of Abbreviations

AIDS	Acquired Immunodeficiency Syndrome
APHRC	African Population and Health Research Centre
ASFR	Age Specific Fertility Rate
CDC	Center for Disease Control and Prevention
CHS	Community Health Stations
DH	District Hospital
DHS	Demographic and Health Survey
DSA	Demographic Surveillance Area
EPI	Expanded Program on Immunization
HDSS	Health and Demographic Surveillance System
HIV	Human Immunodeficiency Virus
HR	Hazard Ratio
IMCI	Integrated Management of Childhood Illness
IV	Instrumental variable
KEMRI	Kenya Medical Research Institute
KHDSS	Kisumu Health and Demographic Surveillance System
KIDS	KwaZulu-Natal Income Dynamics Study
KPLC	Kenya Power and Lighting Company
LTFU	Loss To Follow Up
MDG	Millennium Development Goal
NELM	New Economics of Labour Migration
NGO	Non Governmental Organization
NUHDSS	Nairobi Urban Health and Demographic Surveillance System
OLS	Ordinary Least Squares
PCA	Principal Component Analysis
PHC	Primary Health Care
PSLSD	the Project for Statistics on Living Standards and Development
SES	Socio-Economic Status
TFR	Total Fertility Rate
UDHS	Ugandan Demographic and Health Survey
WHO	World Health Organization

Notes on Contributors

Kubaje Adazu chief demographer and technical consultant of the Health Demographic Surveillance System (HDSS) unit for the Centres of Disease Control (CDC) in Kisumu, Kenya, passed away on Monday, 19 January 2009 at his home in Kisumu.

Adazu was co-leader of the INDEPTH Migration and Urbanization Working Group. He started his career in 1992 as a data manager at the Navrongo Health Research Centre in northern Ghana, which runs the Navrongo HDSS. He left Navrongo in 1997 to go to the US to pursue his studies. He graduated from Brown University in 1999 with an MA in Sociology and in 2002 with a PhD in Sociology. He proceeded to the Kisumu HDSS in Kenya where he was a consultant for the CDC. He provided technical support in the design and launch of the HDSS in Kisumu. From December 2003 to November 2005, he was ORISE Fellow at CDC Kenya where he coordinated research of the HDSS to determine the effectiveness of control programs for diseases of public health importance. He was promoted in December 2005 to chief demographer and technical consultant for the Demographic Surveillance System unit for the CDC/KEMRI Field Station where he was working until the time of his death. His brief included assessing the contribution of demographic factors such as migration, fertility and population age structure on morbidity and mortality due to diseases of major public health importance. Dr Adazu has widely published in referable technical publications and won numerous awards and fellowships from various organizations, including Mellon, Hewlett, Compton Foundations, Population Reference Bureau and Population Council.

Nurul Alam is a statistician and did his MA and PhD in Medical Demography at the London School of Hygiene and Medicine. He has been working with the health and demographic surveillance unit of the Public Health Sciences Division of ICDDR,B since 1980. Dr Alam has participated in a number of national sample surveys and surveillance in different capacities. His research interests include public health and nutrition issues; morbidity, nutrition, mortality and healthcare seeking of women and children and demographic issues; population and development, nuptiality, fertility and migration (national and international) and urbanization. He has attended a number of national and international conferences to present research findings and has also published a considerable number of articles in research journals.

Pedro Alonso is currently the director of the Barcelona Centre for International Health Research (CRESIB) at the Hospital Clinic, professor at the University of Barcelona and Chair of the Board of Governors of Fundaçao Manhiça in Mozambique. His professional career has been focused on the most important global health problems, especially those affecting developing countries. The four main areas of his work are: research, education, medical assistance and international cooperation. The cornerstone of Dr Alonso's research activity has been the development and testing of new control tools against the main infectious diseases such as malaria, HIV, acute respiratory infections and other communicable diseases, specially those tools which help to reduce the morbidity and mortality in less developed countries.

John Aponte graduated in medicine at the Pontificia Universidad Javeriana in Bogotá, Colombia. He obtained an MSc in Medical Statistics from the London School of Hygiene and Tropical Medicine. Dr Aponte has worked on the evaluation of malaria control tools in endemic countries since 1991. He joined the team of Dr Alonso in 1996. Aponte has participated in the design, implementation and analysis of several clinical trials, including the evaluation of prophylaxis against malaria during the first year of life in endemic regions, evaluation of intermittent treatment in infants and the evaluation of malaria vaccines both in Tanzania and in Mozambique. Dr Aponte has also participated in the design of the Demographic Surveillance carried out in the Manhiça Health Research Centre in Mozambique since 1996. Currently, he is the head of the Statistics Unit at the CRESIB, Hospital Clinic in Barcelona.

Donatien Beguy joined APHRC in August 2007 after receiving a PhD in Demography from University – Paris 10 (France) in 2007. He previously worked as a research assistant at the Demographic Research Unit of the University of Lomé, Togo (1998–2002). His research interests include economic demography, transition to adulthood, migration and urbanization. He is a member of many professional associations including Demography Network of Francophone University Agency, Union for African Population Studies (UAPS), International Union of Scientific Study of Population (IUSSP) and Population Association of America (PAA).

Philippe Bocquier is a social demographer and has been working in and on West and East Africa for almost 20 years collecting and analysing mainly longitudinal data. He specialized in the fields of urbanization and urban projections, migration and urban integration and migration and health. Formerly research fellow at Institut de Recherche pour le Développement, he is now senior lecturer at the University of the Witwatersrand.

Nguyen Thi Kim Chuc, pharmacist, MSc, PhD, is an associate professor and vice director of Family Medicine Department at Hanoi Medical University. Dr Chuc has been involved in medical research and education since 1990. Her field of expertise is the assessment of health care system and medical education systems and the implementation of national and international research in the health sector, particularly in the area of primary health care. In this capacity she has been the coordinator of several projects such as the Health System Research Project supported by Sida/SAREC, family medicine development in Vietnam, supported by CMB and many others. Dr Chuc is also the leader of the Filabavi Demographic Surveillance Site. She has many publications on different topics published in research journals internationally.

Samuel J. Clark is a demographer and an assistant professor in the Department of Sociology at the University of Washington. He is trained in biology, computer science and demography and his research focuses on issues that affect Africa. Professor Clark's recent work has pursued simulation-based studies of the impact of HIV on African populations, methods development to improve the value of estimated and modelled results, empirical investigation of migration and mortality in southern Africa, methods to improve the management and analysis of longitudinal population data and capacity development for population and health research in Africa and Asia.

Mark A. Collinson is a health and population researcher and field research manager for the Medical Research Council/Wits University Rural Public Health and Health Transitions Research Unit, School of Public Health, University of the Witwatersrand, South Africa. He is a guest researcher at the Centre for Global Health Research, Umeå University, Sweden. He is responsible for the Agincourt Health and Demographic Surveillance System (HDSS) and oversees the field and data aspects of research embedded in the HDSS. As a researcher, Collinson leads the scientific theme, 'Households Responding to Shocks and Stresses', which comprises a trans-disciplinary network of scientists examining the dynamics of livelihoods, migration, social connection and health. Collinson also convenes the Migration and Urbanization Working Group of the INDEPTH Network. He obtained an MSc in Medicine from the University of the Witwatersrand and a PhD in Epidemiology and Public Health from Umeå University in Sweden.

Daniel Feikin has been at the Centres for Disease Control (CDC) for twelve years. He currently serves as the epidemiology team leader for CDC's International Emerging Infections Program in Kenya. He manages a team of 30 staff running a population-based surveillance project in western Kenya. He is currently a co-principal investigator (PI) on a multi-centre randomized controlled trial of rotavirus vaccine. He previously ran the global health project portfolio of the Respiratory Diseases Branch at CDC. He has coordinated projects that have assessed the efficacy and impact of the Hib conjugate vaccine in several African countries, developed protocols to measure the burden of Hib and pneumococcal disease with the WHO and worked as the lead investigator of a project evaluating the pneumococcal conjugate vaccine in HIV-infected adults.

Sally E. Findley is a professor of Clinical Population and Family Health (in Paediatrics) at the Mailman School of Public Health, Columbia University. Trained as a sociologist and demographer at Brown University, she has long conducted research on international, rural-urban migration and rural development. In 1977 she authored *Planning for Internal Migration*, which set the tone for a pragmatic view of the relation between migration and development policies. She also pioneered the use of contextual models to assess multiple influences on migration decision-making. She completed a post-doctoral fellowship on drought survival strategies in the Sahel, through the Sahel Institute in Mali and since then has devoted much of her research career to migration trends in Sub-Saharan Africa. She is the author of numerous papers and chapters concerning women and migration in Africa, drought and migration and African migration trends. She currently leads a study of the relation of climate to migration and child health in Mali. Her advisor roles include advisor to the Migration and Urbanization Working Group of the INDEPTH Network of surveillance sites and the Global AIDS program of the Centres of Disease Control. In addition, she is a member of the Executive Committee of the Columbia University Institute for African Studies.

Annette A.M. Gerritsen was professionally trained in epidemiology in The Netherlands. She obtained her PhD from the Vrije Universiteit in Amsterdam, reporting on a randomized controlled clinical trial on the treatment of carpal tunnel syndrome. After that she has been involved in research studies focusing on the health and health care use of refugees, asylum seekers and other migrants in the Netherlands. In 2007 she moved to South Africa to work as a senior lecturer in Epidemiology and Biostatistics at the University of Venda, Thohoyandou, where she is the coordinator of the Master of Public Health program. Dr Gerritsen is involved in several projects on infectious diseases (malaria HIV/AIDS) and migration and health, with her main interest being the research methodology. She is the founder of Epi Results, a consultancy firm which specializes in research and training in epidemiology and quantitative research methods.

Philip Guest is assistant director of the Population Division at the United Nations where he serves as the chief of the Demographic Analysis Section. Prior to joining the United Nations he worked as an associate professor at Mahidol University and as a senior consulting associate at the Population Council. He has undertaken research in Cambodia, China, Lao PDR, Mongolia, Myanmar, Thailand and Vietnam. His research interests include reproductive health, interventions research, gender and migration. Dr Guest has worked at the Australian National University, University of Washington and Cornell University. He graduated from the University of Tasmania and holds an MA and a PhD in sociology from Brown University.

Kathleen Kahn is senior researcher in the MRC/Wits Rural Public Health and Health Transitions Research Unit (Agincourt) and senior lecturer in the School of Public Health, University of the Witwatersrand. She has a medical and public health background. Kahn leads the Agincourt Unit's work on child health and development, as well as work on mortality, in particular determination of cause-of-death based on a validated verbal autopsy instrument. Kahn heads the unit's growing Graduate Research Training Programme. She has co-authored over 35 publications. She is an honorary senior lecturer in the Health Promotion Research Unit of the London School of Hygiene and Tropical Medicine, UK, a research associate in the Population Centre, University of Colorado at Boulder and guest researcher in Epidemiology and Public Health Sciences at Umeå University, Sweden.

Rose Kiriinya is currently a project data analyst for CDC/Emory University Kenya Health Workforce Project (KHWFP). Before joining KHWFP in June 2008, she was a research officer at the KEMRI/CDC Health and Demographic Surveillance System (KHDSS) program in western Kenya for five years. Her research interests include child mortality, adolescent health and contribution of human behaviour in accessing health services among populations in sub-Saharan Africa. She holds an MSc in Population-based Field Epidemiology from the University of the Witwatersrand in Johannesburg, South Africa and a BA (Mathematics) degree from the University of Nairobi, Kenya. In addition, Rose has been trained in cultural epidemiology with the Swiss Tropical Institute. Previously, she helped in the analysis and documentation of the UNICEF study on 'South Sudan Basic Education Sub sector Analysis' and 'Analysis of Status of Education in Southern Sudan', a study by Secretariat of Education and Demobilization of Child Soldiers (SPLM) and Catholic Relief Sudan (CRS).

Adama Konseiga joined the Institute for Development in Economics and Administration (IDEA International Institute, Quebec City) in September 2007. He is also a research affiliate of the GREDI (Research Group in Economics and International Development) of the University of Sherbrooke, Canada since May 2007. He worked at the Department of Economics of the Faculty of Administration as a lecturer. He is also a research fellow of the Institute for the Study of Labour (IZA, Bonn).

Before joining the current position, he held a position at the African Population and Health Research Centre (APHRC) as a post doctoral research fellow from August 2005 to April 2007. In 2005, he held a short-term research fellow position at the Department of Economic and Management Sciences, University of Pretoria, South Africa. He has previously worked at the Ministry of Economic Development and Finance as a project manager in charge of the European Development Fund Projects for Burkina Faso. Dr Konseiga received a PhD in Economics in 2004 as a unique collaboration between the University of Bonn in Germany and the University of Auvergne in France. He also holds an MA in Development Economics from the University of Auvergne (at CERDI, Centre d'Etudes et de Recherches sur le Développement International in France). His areas of research interest include urban poverty, migration, education, health and economic convergence.

Kayla Laserson has been at CDC for eleven years. Currently she is the director of the KEMRI/CDC Field Research Station where she oversees a staff of over 950 individuals and a comprehensive research program in HIV, malaria, TB, emerging infectious diseases, demographic surveillance and programmatic service delivery of HIV care, treatment and prevention programs. She is the overall principal investigator of the KEMRI/CDC Health and Demographic Surveillance System and a co-PI on a phase three randomized placebo-controlled multi-centre trial to evaluate the efficacy and safety of a rotavirus vaccine, a multi-centre phase three malaria vaccine trial and two cohort studies, one among infants and the other among adolescents, to prepare the KEMRI/CDC demographic surveillance area for the advent of phase three tuberculosis vaccine trials. Prior to coming to Kenya, she worked as the deputy chief of the International TB Branch in the Division of TB Elimination at CDC-Atlanta. There she supervised a team of medical epidemiologists in the provision of international program support to global tuberculosis programs, in particular to those countries providing the largest number of immigrants with tuberculosis to the US and those contributing the most to the global burden of tuberculosis. She was responsible for the design and conduct of international tuberculosis investigations, including all epidemiological analysis and conducted operations research training to build capacity to facilitate operational research to improve tuberculosis program performance and reduce the global tuberculosis burden.

Leonildo Matsinhe is a geographer working at Manhiça Health Research Centre as a training fellow in demography. His work involves supervision of field activities including data collection procedures, data cleaning and management in Manhiça HDSS. His research focus is on vital events including births, deaths and migration and other data such as pregnancies, abortions, stillbirths and level of education.

Cheikh Mbacké is senior advisor to the William and Flora Hewlett Foundation's Population Program since July 2006. In this capacity, he works in consultation with the Foundation's officers to help improve the training of African population scientists and to increase the availability of population, reproductive health and related data to scholars in Africa. Before taking this assignment, Dr Mbacké worked for the Rockefeller Foundation for fourteen years occupying successively the positions of program officer for population sciences (1992–1999), representative for eastern Africa (1999–2000), director of the Africa Regional Program based in Nairobi, Kenya (2000–2003), vice president for Administration and Regional Programs based in the Foundation's headquarters in New York City (2003–2005) and senior advisor working from Dakar, Senegal until June 2006. Dr Mbacké, a statistician and population scientist by training, began his career as a civil servant at the Senegalese Statistical Bureau in January 1976 before spending six years as a researcher and head of the training division of the Centre for Applied Studies and Research on Population and Development (CERPOD) at the Sahel Institute in Bamako, Mali from 1986–1992. Dr Mbacké has a BA degree in Statistics from the Institut National de la Statistique et des Etudes Economiques (INSEE) in Paris, France, 1975; an MA degree in Demography from the Institut de Formation et de Recherche Démographique (IFORD) in Yaoundé, Cameroon, 1979 and a PhD in Demography from the University of Pennsylvania in Philadelphia, Pennsylvania, US, 1986.

Kanyiva Muindi first joined APHRC in 2000, working on the Nairobi Urban Health and Demographic Surveillance System (NUHDSS) as a field staff member. She later proceeded to do an MSc in Public Health/Epidemiology and Biostatistics at the University of the Witwatersrand, Johannesburg and rejoined the Centre in August 2006. She is currently working as a research officer on the Urbanization, Poverty and Health Dynamics in Sub-Saharan Africa program (UPHD) specifically on the migration, poverty and transitions to adulthood projects. Her research interests include migration, adolescent sexuality and environmental health.

Ariel Nhacolo is currently the head of the Department of Demography at the Manhiça Health Research Centre (CISM) in Mozambique, where he has been working since 1999 as a training fellow in medical demography. During the last three years he was the coordinator of the CISM. His background is in geography and he has an MSc in Demography and Health. His research focus has been on all aspects of pure and social demography mostly those related to fertility, mortality and migration among adults in rural settings.

Delino Nhalungo is a geographer and leads the Health and Demographic Surveillance System (HDSS) data management. His work is oriented on development of tools for demographic data collection and management, as well as data analysis, with emphasis on special variations. His research is on the domain of levels and trends of mortality and the impact of migration in health.

David Obor has been working at the KEMRI/CDC KHDSS for five years. Currently he is the data manager and analyst and oversees over 13 data processing staff. In this responsibility he leads the team's processing, quality maintenance and analysis of the KHDSS data. The KHDSS population-based longitudinal data is critical in the design and measurement of appropriate health, social, economic, behavioural interventions for the study area (for example bed nets, ARVs, vaccines). He holds a BSc from Nairobi University, 1996.

Peter Ofware worked with KEMRI/CDC for seven years up to December 2008. Currently he is the program manager for Child and Reproductive Health and focal person for Community Partnering for AMREF in Kenya. As program manager, he provides strategic leadership in the development of the program across the

AMREF country program. He oversees a staff of over 50 individuals working in seven different projects across the country, implementing maternal and child health, adolescent and youth reproductive health and PMCTC activities. He is the focal person for community partnering within AMREF in Kenya which is in line with AMREF's ten year strategic plan of putting African communities first. While working with KEMRI/CDC he was instrumental in setting up the HDSS in 2001 and served as the HDSS field coordinator from 2004–2007. In early 2007, he coordinated and implemented the expansion of the HDSS into a new area, namely Karemo Division, now a key area of the HDSS where current clinical trials are being conducted. Over the years, he has coordinated numerous sub-study surveys in the HDSS, including annual data collection of malaria anaemia and parasitemia data and collection of self-reported HIV data. In 2008 he served as the special studies coordinator for the HDSS and helped to coordinate the plans for the expansion of home-based HIV testing to the rest of the HDSS area, greater immunization data collection and data collection among the internally displaced persons who were residing in the HDSS.

Bernard Onyango is currently a graduate student at Brown University in the social demography program that is nested within the PhD program in Sociology. His research interests include mortality, education, migration and fertility with particular focus on Sub-Saharan Africa. Before joining Brown University in 2007, Bernard was a research assistant at the KEMRI/CDC KHDSS program in western Kenya for five years. He holds a BA degree (2001) from the University of Nairobi.

Ho Dang Phuc, PhD, is a mathematician. Since 2006 Dr Phuc has been the head of the Department of Probability and Mathematical Statistics at the Institute of Mathematics, Vietnamese Academy of Sciences and Technology. His fields of expertise are theory of limit theorems in probability and statistical data analysis. He collaborates with the Filabavi Demographic Surveillance Site as a main statistician of the site and is involved in several studies conducted by that field laboratory.

Sureeporn Punpuing is an associate professor and director at the Institute for Population and Social Research (IPSR), Mahidol University. Her research focuses primarily on cross-border migration, impacts of migration on the family and elderly left behind and population-environment interactions and health. She was the site leader of the Kanchanaburi Demographic Surveillance System (DSS) and an elected board member of the INDEPTH Network. Dr Punpuing has worked as a population affairs officer at the United Nations in New York, a consultant for ESCAP and IOM in Bangkok and UNFPA in Mongolia. Dr Punpuing teaches advanced analysis of migration and human ecology to MA and PhD students. She holds a BA in Statistics in Thailand, an MA in Demography and a PhD in Resource and Environmental Studies from the Australian National University, Australia.

Charfudin Sacoor is a population and demography researcher for the CISM, Mozambique. He supervises field activities including planning and data collection procedures, data cleaning and management for the Manhiça HDSS. He has also participated in other demographic components of several surveys carried out in Manhiça including the malaria vaccine trial. His background is in geography and currently he is a post-graduate student at Wits University. His recent research has been on demographic indices and trends as well as under 15 year's mortality.

Laurence Slutsker is chief of the Malaria Branch at CDC and has an appointment as a clinical assistant professor of Medicine, Emory University School of Medicine, Atlanta, Georgia. Dr Slutsker received his BSc from the University of Michigan and his medical degree from Case Western Reserve University in Cleveland, Ohio. In addition, he also holds an MA in Public Health from the University of California, Berkeley. Dr Slutsker completed his residency in Internal Medicine at the University of North Carolina, Chapel Hill and is board certified in both internal medicine and preventive medicine. In 1987, Dr Slutsker joined the Malaria Branch at CDC as an epidemic intelligence service officer. In addition to 14 years experience with the Malaria Branch, Dr Slutsker has also held staff positions in HIV/AIDS and diarrhoeal diseases and has conducted epidemiologic research in a variety of areas including causes of infant mortality in developing countries, HIV/AIDS, malaria, diarrhoeal diseases and general tropical public health. From 2001–2005, Dr Slutsker was the director of the CDC/KEMRI Research Station in western Kenya. In 2005 he returned to Atlanta to become Chief of the Malaria Branch at CDC. His current research interests include: prevention of malaria in infants and pregnant women, malaria/HIV interactions, anti-malarial drug resistance and malaria vaccine evaluations.

Dr Slutsker has conducted and supervised epidemiologic research in the United States, Africa, India and Asia. He has lectured widely in the US and abroad on his research and on general public health issues. He has authored or co-authored more than 140 scientific journal articles, book chapters and other publications.

Peter Kim Streatfield has been head of the Matlab Health and Demographic Surveillance System since 1999 and head of the Population Programme at ICDDR,B. Previously he was the country representative of the Population Council in the Bangladesh country office. His background is in physiology and medical demography. His recent research has been on climate and morbidity/mortality; urban health; sexual behaviour and STDs; aging and health; chronic diseases; nutrition; consequences of arsenic in drinking water; health equity; and maternal health.

Nguyen Xuan Thanh is a health economist at the Institute of Health Economics, Edmonton, Canada specializing in health economics, evaluation in public health, biostatistics and epidemiology. He received training as a medical doctor at Hanoi Medical University in Vietnam and holds both an MA and PhD in Public Health from the Umea International School of Public Health in Sweden. Dr Thanh has a wide range of experience working as a consultant for the World Health Organization, World Bank and the Ministry of Health of Vietnam among other organizations. He has published many journal articles, co-authored several books and singly authored a book on the injury poverty trap in rural Vietnam.

Stephen M. Tollman, BSc, MMed (Wits), MPH (Harvard), MA (Oxon), PhD (Umeå) is based in the University of the Witwatersrand, South Africa, where he heads the School of Public Health's Health and Population Division and chairs the university's Population Program Executive and Forced Migration Steering Committee. He directs the Medical Research Council/University Unit in Rural Public Health and Health Transitions Research. Professor Tollman currently chairs the INDEPTH Board of Trustees and is principal investigator for its initiative in Adult Health and Aging. A Rhodes Scholar, Tollman recently led an external evaluation of the Alliance for Health Policy and Systems Research, an initiative of the Global Forum for Health Research.

John Vulule has been at KEMRI for the last 18 years. Currently he is a chief research officer and director for the Centre for Global Health Research of KEMRI based in Kisumu. He oversees a work force of over 1,100 individuals and extensive research programs in HIV, malaria, emerging infectious diseases, demographic surveillance and programmatic service delivery of HIV care, treatment and prevention programs. He is the principal investigator (PI) of the KEMRI/CDC Cooperative Agreement and Co-PI in several other studies. Prior to his current duties, he was extensively involved in malaria studies specifically disease control using insecticide treated bed nets. He has both supervised and examined several PhD and MSc students and is also a part-time lecturer in a number of Kenyan universities.

Michael J. White serves on the scientific advisory committee of the Migration and Urbanization Working Group for the INDEPTH Network. Professor White has been on the faculty of Brown University since 1989. He is professor of Sociology at Brown and currently director of the Population Studies and Training Centre. He received his PhD in Sociology in 1980 and has held appointments at Princeton University and the Urban Institute, Washington DC. Professor White served as a member of the US National Academy of Sciences Panel on Urban Population Dynamics and he has been a fellow of the Woodrow Wilson International Centre for Scholars. He has also served on the board of directors of the Population Association of America, the NICHD Population Studies Committee and he is recent past president of the Association of Population Centres. Professor White's research interests span a broad range of demographic topics with a focus on population distribution, urbanization and migration. He maintains active research collaboration on these topics in a variety of developed and developing settings.

Yazoumé Yé has undergraduate training in Geography from the University of Ouagadougou. He joined the APHRC in May 2006 after completing his PhD in Public Health/Epidemiology at the Medical Faculty of the University of Heidelberg, Germany. Yazoumé also holds an MSc in Community Health and Health Management in Developing Countries which he obtained in 2002 from the same university. Prior to that (1996–2001) he held a position at the Nouna Health Research Centre as database developer and manager and head of the Department of Information Management. Before that (1996–1995), he was a high school lecturer in geography and history at the Po High School in Burkina Faso. His research interests include malaria epidemiology (modelling transmission risk), physical environment and health, perceived quality of health care, urbanization and health.

Eliya M. Zulu is the deputy director and director of research at the APHRC. He is a demographer with a PhD in Demography from the University of Pennsylvania, USA (1996) and an MA in Population and Development from the Australian National University (1991). Dr Zulu joined APHRC in 1997. Previously he worked at the Demographic Unit, University of Malawi, as a lecturer in demography prior to his doctoral studies. His research centres on linkages between urbanization, poverty and health outcomes, sexual and reproductive health (particularly among adolescents), schooling and bridging the gap between research, policy and action. He has wide experience in research among the urban poor in sub-Saharan Africa and has published extensively on population and health issues, including sexual networking and maternal health in sub-Saharan Africa. He played an instrumental role in designing and is currently managing the longitudinal Health and Demographic Surveillance System that APHRC is implementing in two slum settlements in Nairobi City. Dr Zulu has served on many international technical panels for improving data systems in Africa and has led a number of multi-partner and multi-national research projects on reproductive health and urban poverty and health outcomes. He also serves on editorial boards for two leading journals in the population field.

Dr Zulu is the current president of the Union of African Population Studies (UAPS), a pan-African non profit Scientific Organization.

Foreword

While not a scholar of migration I have extensive personal awareness of this having grown up in the rural Transkei where migration was part of the normal life-course for young men. It was a recognized rite of passage for youths who would depart for the mines or cities ending up in places like Cape Town or Johannesburg. Some would return and others not. Those who returned were known as 'amagoduka' or 'those who came back' and those who stayed on were called 'amathsipha' which means those who stayed. For the 'amagoduka' their accommodation was mostly in single-sex hostels. Cape Town was a 'Coloured preferential area' and there was a lot of divide-and-rule between 'amagoduka' and the township dwellers who enjoyed Section 10(b) 'privileges',[1] and between the 'Coloured' and people of African ancestry. To this day these hostels are still used to house workers and their families.

People hold on to their rural identities regardless of how long they spend away and it is normal to want to go back and die and be buried in the home area. Even for established professionals who have lived in the city most of their lives the idea of being buried there, and not where past generations have lived, contradicts a core identity. Letting go of the rural linkage, of whatever form, seems to signify turning one's back on one's ancestors. There is a form of spiritual linkage, a sense of belonging, even in cases where people had Section 10(b) rights. Historical rights go back a long way and, interestingly, the pull is not always in the direction of returning from the city. When I studied in the Western Cape I encountered people coming from the Transkei to claim their rights to live in Cape Town. This example highlights not just the just the role of identity driving migration but also the violence of a pre-democratic era when migration was imposed by the government in pursuit of racially tidy urban areas. Since returning home is an important part of the normal trajectory of migration it is important to think of the right of return, and look at situations where people have lost this right.

This brings up the politics of internal oppression which makes migration a subject relevant to all South Africans. Rural development is a priority for the present government with infra-structure development a key component. Gaining access to facilities and schools is one of the driving forces moving people out of rural areas towards situations where there are more opportunities and people can live closer to their place of work. Even the classification of rural and urban areas in South Africa is marked by the differential in access to facilities and schools rather than whether or not people conduct agriculture. Tackling these ongoing

1 Section 10 rights meant to have the right to remain permanently in cities.

inequities is part of dealing with the legacy of apartheid. An important aspect is the availability of housing. People need a decent place to live, whether in a rural or urban area, and this can be a key part of a decision to migrate.

In the past, but increasingly since democratization, there is also a large volume of female migration. Some is linked to whether or not the husband is a migrant, but often not. The absence of the male in the family is more accepted and less consequential on the smooth running of a household. The absence of the female migrant is harder for children and more so if the family's support structure has been weakened either by death or out-migration of other family members.

Overall, the consequences of migration for traditional rural communities have been complex and this makes it a good subject for research in some of the world's poorer countries. This study brings together contributions from INDEPTH scientists which are unique for the type of data used. This has been meticulously compiled from many thousands of visits to households in rural areas or urban slums in Africa and Asia. *The Dynamics of Migration, Livelihoods and Health* demonstrates the importance of comparative research. While global intellectual resources and new data can bring insight into this problem wherever it is studied, this book has the added benefit of the study being driven by Southern hemisphere scientists. Their efforts are striving to offset global inequities, not unlike the case of South African authorities investing in marginalised communities as a development initiative. Strong research programmes in 'Southern' universities can help to stem the much maligned brain-drain by retaining intellectual resources in countries that need them most and where the research findings are highly needed.

Professor Loyiso Nongxa
Vice Chancellor, University of the Witwatersrand, Johannesburg

Preface

For years at INDEPTH sites, counting migrants has been viewed as part of the demographic accounting required to know exactly who is living in a community. The Migration and Urbanization Working Group (MUWG) was established by a group of INDEPTH researchers who believed that migration was much more than that, convinced that understanding migration was key to some of the most critical issues communities face. This book is a result of that collaboration.

Migration is defined as people moving into and out of the HDSS study sites. The book therefore presents a different view from most other migration literature due to the different data structure employed in the analyses. Features of the data include the ongoing registration of a whole district population (or an area of similar size) so that accurate rates can be computed, based on verified moves that occurred in the population. Longitudinal analytic methods, such as event history analysis, pay special attention to the exposures related to migration.

Migration plays a key role in the shaping of family livelihoods and well-being, but discerning those linkages has proved difficult, precisely because of the dynamic nature of migration, which involves people moving in and out of households over widely varying time frames. There is a scarcity of datasets adequately equipped to examine these dynamic relations. Furthermore, different parts of the world have emphasized different aspects of migration. Countries experiencing high immigration are concerned with the impact of international migration and the adaptation of in-migrants in host communities. Internal migration is more of an issue in developing countries where the proportion of rural population is larger and the economic transition associated with internal migration is higher. But, it is precisely these poorer countries where the least data is available to examine internal migration. New data and scholarship are thus urgently needed on internal migration in developing countries. This book focuses on internal migration and how it impacts on health and education patterns and the dynamics of poverty at household level. With this we hope to add to the trend that INDEPTH multi-country data provides new insight into population variables.

This volume is a peer-reviewed collection of twelve chapters prepared by the INDEPTH Migration and Urbanization Working Group. The first four are introductory and overview chapters, followed by seven site chapters from Health and Demographic Surveillance System (HDSS) sites and an epilogue. The site chapters are divided into two themes. The first theme is Migration and Livelihoods, containing chapters from two Asian sites (Kanchanaburi and Matlab) and one African site (Agincourt); and the second theme is Migration and Health, with one Asian site (Filabavi) and three African sites (Kisumu, Nairobi and Manhiça).

The settings represent six countries, namely Thailand, Bangladesh, South Africa, Vietnam, Kenya and Mozambique. A diversity of settlement types is included from urban to peri-urban to rural. A range of economic levels is covered however the settings generally represent the impoverished parts of poor countries. To summarize the book structure: after introducing the topic and explaining what we can hope to gain from the surveillance approach to tracking migration, a methods chapter provides a detailed reflection on the HDSS methods and how these pertain to the study. The second chapter also includes a comparative table of migration definitions used in the different sites. The third chapter presents the community contexts, an overview of the study settings and explains how the characteristics of a place help to shape the outcomes of migration. To see who actually migrated, Chapter 4 shows the comparative age–sex migration profiles from the participating sites. The book then proceeds to the site chapters grouped into the two themes where site-by-site evidence is presented on the population, health and socioeconomic consequences of migration.

We hope that you will read this volume with interest and that the research presented here will stimulate a re-examination in your own community about where migration fits into that complex puzzle of health and livelihoods.

Mark Collinson (Convener)
Migration and Urbanization Working Group, INDEPTH Network

Acknowledgements

When the INDEPTH Network established in 2003 the Migration and Urbanization Working Group (MUWG) and the group itself subsequently appointed Mark Collinson (University of the Witwatersrand, Agincourt HDSS, South Africa) and Kubaje Adazu, of blessed memory (Kisumu HDSS in Kenya), to lead the group with the aim of developing/strengthening the capacity in INDEPTH HDSS sites to conduct longitudinal migration studies and produce a multi-country scientific volume, we were not unaware of the huge task ahead. Writing this note of thanks in 2009 should confirm how challenging it has been for the group.

We wish to thank Mark and Adazu (posthumously) for selfless leadership of the group. The editorial committee for this cross-site publication was later expanded to include two leading scholars who accepted to share their expertise with our network scientists. We thank Michael White (Brown University, USA) and Sally Findley (Columbia University, USA).

The MUWG committee has overseen a six-year period of training, data preparation, analysis and write-up to produce this publication. We thank Samuel Clark, Benjamin Clark, Jeffrey Eaton and Peter Byass who provided technical support throughout the period this publication was written.

To assure scientific quality, the INDEPTH usually relies on independent people who review our work and advise our editors. A scientific review panel comprising the following six senior scientists was constituted: Cheikh Mbacké (William and Flora Hewlett Foundation's Population Program, chair); Pierre Ngom (Family Health International, Kenya); Philippe Bocquier (University of the Witwatersrand, South Africa); Rosalia Sciortino (Mahidol University, Thailand); Barney Cohen (National Research Council, USA) and John Oucho (Warwick University, UK). We would also like to thank Annette Gerritsen (University of Venda, South Africa) who came on board as a copyeditor at the last third of the book's production. Annette put much care into the preparation of the manuscript. Other efforts we are grateful for were contributed by Mildred Shabangu (University of the Witwatersrand, Agincourt HDSS, South Africa) and Thando Gwetu (University of Venda, South Africa).

Other colleagues provided institutional support for the work. Noteworthy is the support provided by the Johannesburg-based Health and Population Division, at the University of the Witwatersrand: Stephen Tollman, Kathy Kahn, Sharon Fonn, Dereshni Ramnarain, Zubeida Bagus, Melta Buthelezi, Linda Oldert and Sadiya Ooni, who are a satellite secretariat for the INDEPTH Network in Johannesburg. We need to mention the following three international workshops which were held from 2003–2006 to develop the work presented in the volume:

1. An INDEPTH multi-site migration workshop was held in Johannesburg in January 2003. The meeting was hosted by Stephen Tollman of the Health and Population Division, University of the Witwatersrand. The meeting was funded by the Andrew W. Mellon Foundation. The aim of the conference was to launch the INDEPTH migration and urbanization initiative on a strong scientific footing. Ideas were developed and the group resolved to produce a book. Mark and Adazu were appointed to lead the group. Thirteen INDEPTH sites participated: Agincourt (South Africa), Nairobi (Kenya), BRAC (Bangladesh), Butajira (Ethiopia), Dikgale (South Africa), Filabavi (Vietnam), AMK (Bangladesh), Kanchanaburi (Thailand), Kisumu (Kenya), Manhiça (Mozambique), Navrongo (Ghana), Rakai (Uganda) and Vadu (India).
2. The first INDEPTH MUWG analytic workshop in Kisumu, Kenya was held in November 2004. The late Kubaje Adazu and Bernard Onyango hosted the workshop. The main outcome was to progress the science of the book, including, methodological and theoretical training, database development, training in single episode event history analysis and the production of age–sex migration profiles. Participating INDEPTH sites were: Filabavi (Vietnam), Africa Centre (South Africa), Butajira (Ethiopia), Navrongo (Ghana), AMK (Bangladesh), Nairobi (Kenya), Manhiça (Mozambique), Dikgale (South Africa), Kisumu (Kenya) and Agincourt (South Africa).
3. The second INDEPTH MUWG analytic workshop was held at the University of the Witwatersrand, Johannesburg, 28 August–1 September 2006. The meeting was hosted by Mark Collinson and Mildred Shabangu. We are grateful to the Wits Mellon Migration Node who provided funding for this meeting. The participating sites were: Agincourt (South Africa), Manhiça (Mozambique), Niakhar (Senegal), Navrongo (Ghana), Nairobi (Kenya), Kisumu (Kenya), Kanchanaburi (Thailand) and Filabavi (Vietnam). Advanced statistical input was provided by Michael White (Brown University, USA) and the late Kubaje Adazu (Kisumu, Kenya). Data support was provided by Benjamin Clark (University of the Witwatersrand, Agincourt HDSS, South Africa) and Jeff Eaton (University of Washington, USA).

It is unfortunate, but understandable for products such as this one, that not all INDEPTH site contributions were able to make it into this publication. The number of sites reduced from 13 to seven in the end. We still thank those sites that did not make it, but participated at various stages. We sincerely hope that more sites will be part of further developing the research infrastructure on migration and urbanization within the INDEPTH Network.

During the six-year period in which this work was carried out, the INDEPTH Network received institutional support from the World Bank, Sida/GLOBFORSK, Wellcome Trust, Wits Mellon Node, Rockefeller Foundation, Bill and Melinda Gates Foundation and the Hewlett Foundation. We would like to acknowledge the invaluable contributions of these funding institutions and several others who continue to provide critical funds to the INDEPTH Network to ensure that we complete our work successfully.

Finally, we would like to thank our colleagues at the INDEPTH Secretariat, especially Ayaga Bawah, for their efficient co-ordination.

Osman Sankoh
Executive Director
INDEPTH Network
Accra, Ghana
2009

PART I
Introduction

Chapter 1
Migration and Demographic Surveillance: An Overview of Opportunities and Challenges

Michael J. White

Introduction

Health and Demographic Surveillance Systems provide a unique window onto the process of change in communities. The health and well-being of populations unfold dynamically, often through a sequence of interrelated events. While that is an easy truism, the availability of detailed information to help untangle those interrelationships is difficult to acquire. Temporally detailed data, such as that collected by Health and Demographic Surveillance Systems (HDSS), can help us open a window on this complex interplay of events. By repeatedly observing the same individuals and households in their local communities, we can better understand what factors weigh more heavily on – or predict most strongly – an economic decision or a health outcome. While such surveillance systems were often developed originally with a specific health-intervention or health-monitoring goal in mind, they have expanded recently in scope and so now can provide a resource to help understand demographic change more generally. This volume is the continuation in a series of monographs begun some years ago (Sankoh et al. 2002) that exploit data from the growing set of surveillance sites connected through the INDEPTH network, while focusing on specific demographic and health issues pertinent to low-resource settings.

This first chapter of *The Dynamics of Migration, Health and Livelihoods* gives an overview of the position of migration in HDSS. The chapter sketches key characteristics of a HDSS, as distinguished from other health and demographic data collection styles. The chapter then focuses more specifically on the event of migration, including more specific discussion of the value of HDSS data for understanding the relation of migration to livelihoods and to health, the two thematic applications of this volume. Within each of these two sub-sections, we give brief overviews of the contributions of individual chapters from the several HDSS contributing to this book. The conclusion of this chapter offers some overall points about the prospect of analysing migration and its links to other demographic and social phenomena, employing HDSS.

The remainder of this introductory section of the book is devoted to methodological and cross-site discussion. Chapter 2 focuses in some detail on migration data and methods for surveillance systems. While the first chapter alludes to some of the benefits and challenges of employing migration data in such prospective data collection, Chapter 2 presents these matters with greater conceptual background and makes specific methodological points. Chapter 3 discusses the community context of migration. Migration and their households are embedded within communities. Community conditions, physical, economic and social, play a role in determining migration, both departure and return, so this chapter help set the stage with discussion of community context overall and presentation of comparative data for each site. Chapter 4, the final chapter of the introductory section, presents actual comparative tabulations – age–sex profiles – of migration for the several HDSS sites included in the volume. These data show the basic comparability of migratory behaviour in low resource settings, while also highlighting some potential differences. Taken together, these four introductory chapters are designed to provide a conceptual and methodological framework to inform and illuminate the topical presentation of the seven site chapters to follow.

Migration and Surveillance

Central to the management of a demographic surveillance system is a proper understanding of the migratory behaviour of the population. This is important, crucial in fact, for several reasons. First no matter the scientific and policy objective of the particular HDSS, people inevitably enter and leave the surveillance setting, thereby influencing the ability to properly represent the population under study in any calculated statistics. Second, the behaviour of migration may itself be related to the demographic and health outcomes of the HDSS population, say through contributing remittances from work in another region or being a source of disease risk or new knowledge as migrants circulate among communities. Finally and related to this point, the migrants themselves may be of interest and following them over time may be particularly challenging to a HDSS system.

In their original implementation most surveillance systems were not designed to take detailed account of migration. Initially, many HDSS were seen as loci for moderate-duration medical interventions, in which the departure of subjects (say children receiving a treatment) was assumed to be unlikely or non-problematic. The majority of HDSS have been established in rural areas, where perhaps the designers also thought the population was relatively sedentary.

If the objective of the surveillance system is to ensure that all are accounted for in the surveillance, migration may be seen as a problem (often a data-management problem) to be resolved. Thus, most HDSS have treated departures (out-migration) as loss-to-follow-up (LTFU). Generally, HDSS residents were not followed to other regions of the country, to urban areas or even to nearby districts beyond the

Demographic Surveillance Area (DSA) boundary. Such a decision might make sense when a short-term intervention (childhood vaccination; bednet trial) is the health outcome of interest. In such cases the population under study may not differ much by who stays and who leaves: children who remain in the HDSS may exhibit much the same immunologic response to a vaccination to those who move away. Moreover, statistical techniques are now standard for handling LTFU: censoring is readily managed in hazard models and other longitudinal approaches.

But this practice of waving away migratory behaviour can only go so far. As HDSS data collection (and some of the scientific analyses embedded within HDSS) goes on for longer times, ignoring migration or treating it a random censoring becomes more problematic. And what is more, when one is intrinsically interested in migration as a demographic, social and health phenomenon, ignoring population movement or treating it as random censoring event becomes simply unacceptable. Consider migration and livelihoods. The decision of a person to migrate (or of a family to strategize to send a migrant to another locale) is likely to have a significant impact on the livelihood of the original family. There may be a loss of agricultural production or child supervision or conversely, there may be a new flow of remittances. In either case, ignoring migration is likely to interfere with one's ability to analyse the dynamics of household well-being. This is the kind of combination of technical and substantive concerns the present volume is designed to address.

If we consider migration from this vantage point, as a critical event in and of itself, the ability to track migration presents an unheralded opportunity to exploit some of the unique advantages of HDSS. Given that HDSS are prospective, temporally dense and comprehensive demographic observation platforms, they possess some distinct advantages over other means by which social and health scientists learn of and integrate population movement into substantive analyses. Census data, advantaged by (usually) nationwide coverage, are decidedly cross-sectional and, if anything, usually ask only about place of birth or residence in a prior year. The Demographic and Health Surveys, long-established nationally representative surveys for low-income countries, offer detailed data on selected topics, but have rather limited information about migration. The two widely used sources of information about populations have considerable utility for science and policy, but neither is prospective. Neither censuses nor many existing surveys can provide needed detail about migration. Some retrospective migration data do exist. At one early point the incorporation of an experimental five-year calendar did provide a window on the potential utility of such surveys for longitudinal migration analysis (White et al. 1995). There are also efforts to employ longer multi-domain event history calendars in developing settings (White et al. 2008). The HDSS offer a distinct contrast. As prospective data collection efforts, the HDSS are privileged in being able to provide information about migrants' departure and return, as well as characteristics of the migrant's household and community before and after.

Migration and Livelihoods

Migration is centrally linked to shifts in the fortunes of individuals and households. Much of this is in turn linked to labour market outcomes, the success or failure of migrants in their new location. Theories related to this behaviour – and the empirical regularities that accord with it – provide an important backdrop for the continuing operation of HDSS. Understanding the social process of migration, as well as its links to health, can help understand who is moving, why and the data management and supplemental survey techniques best equipped to reflect that movement.

The relationship between migration, human capital and labour market outcomes is voluminous and a number of reviews provide a ready introduction to the literature (Brettell and Hollifield 2000, Lucas 1997, Montgomery et al. 2003, White and Lindstrom 2005). This literature includes a substantial treatment of migration in developing countries and rural-urban migration, topics relevant to our HDSS settings in this volume. At the same time, a review of the literature would point to a deficiency in the availability of high-quality longitudinal data with relevant covariates for studying migration in the less developed setting. Typically inferences about migration in Asia and Africa are made from censuses or broad cross-sectional surveys (such as the Demographic and Health Surveys or the Living Standards and Measurement Surveys), which have broad population coverage but limited information about timing or residential histories.

The behavioural links between migration and social change are manifold. We point out some key research topics here, concentrating on phenomena for which HDSS may be particularly well-suited.

The Migrant, the Household and Risk Management

Migration studies have moved well beyond the concept of the individual, atomized migration aiming to maximize a wage across potential destination labour markets. Rather considerable effort, both theoretical and empirical, over recent years has gone into conceptualizing the event of migration in context. In many contemporary models, often gathered under the rubric of the New Economics of Labour Migration (Taylor 1999), migrants are seen as part of a joint decision-making process that manages risk across household members and sees remittances as a conscious household strategy.

Migration of selected household members enables rural households to manage risk. They do so by diversifying sources of income across locations. In a prototypical case, the rural household sends one of its members (say a young unmarried child) to the city to earn wages. Since urban and rural economic fortunes may move in different directions over time, the urban income stream offers a counterbalance to the rural income stream. In this way, not all of the household's income is derived from the rural (often farm) economy, which is subject to whatever may befall the local economy through happenstance of weather and the like. Within this strategy, the household may actually elect to receive lower expected income, but does so precisely because it is managing risk and is less subject to debilitating shortfalls in income from local economic shocks.

While the risk-management paradigm is now well-established in the discourse on migration, less fully developed is the exact process of carrying through the strategy. To phrase it as a question faced by the household: *Who goes where when?* The traits and composition of the household undoubtedly provide clues as to the answer to that question. The size, intra-familial relationships and age of household members must play a role. In some settings, a late adolescent child might be the migrant; in others settings, where no such age-eligible person resides, the household head may migrate. The scale of rural activity (success of the farming enterprise) would be likely to influence who migrates when.

Information also matters. In places where migration, especially circular migration, is a less well developed pattern, there may be less information about opportunities elsewhere. As a consequence, out-migration from the household may be followed before long by return migration by the unsuccessful migrant. Conversely, households residing in communities with considerable circular migration experience can rely on earlier migrants – networks of kin, friends and neighbours – to relay more accurate information. Education of household members, their socioeconomic status (literacy, access to media) may also influence the amount of information coming into the household and the probability of migratory activity.

Finally, it is worth being mindful of the fact that not all migration is for pecuniary gain and wage labour. Members of the household may migrate as part of a wider kin network of support, say to care for an ailing member of the extended network. Migration in order to be nearer care may figure in the pattern and migration 'home' to retire or to receive care when terminally ill, can generate flows of movement as well.

Remittances

Scholars have begun to document the continuing connection between migration origins and destinations. It is not often the case that the migrant cuts ties completely with the origin community and family; rather, repeated visits over the course of a year are likely to take place. Moreover, remittances (or other resources) are likely to keep the migrant in touch with his or her origin household or family (Lindstrom 1996, Cai 2003, Van Wey 2004, Van Wey et al. 2005). These monetary flows presumably improve the well-being of the origin family and the wider community, as the funds circulate. Despite some efforts to document the flows of persons and funds, both within and between countries, our level of knowledge is quite deficient regarding their effects. HDSS offer a promising opportunity to see such effects – by linking the migrant individual and remittance back to the origin over time – on origin household well-being. These financial flows may emerge as health improvements as the recipient household has more resources for food, shelter, health care and the like.

Overall SES

While remittances are seen as an economic benefit of migration, the actual picture is more complicated and the net socioeconomic outcomes less predictable. Remitting migrants surely generate a monetary flow to their home family and community. At the same time, they remove a worker and caretaker from the household. Some would hypothesize, then, that migration may also have adverse consequences for the sending household (Adazu 2002). Thus, the *net* effects of migration for the sending household should be seen as an empirical question. To be sure, the likely impact depends on who leaves. Departure of the breadwinner or separation of a mother from her young children may have deleterious consequences, while sending a young adult to the city to work in industry may be beneficial.

Migration as a risk management strategy has seen some empirical analysis (Sana and Massey 2007, Taylor 1999), but this work remains limited, especially for the lowest income settings. HDSS would be well suited, given the continuous measurement of household composition and the indicators of household well being available – to shed a bright light on this risk-minimization proposition and how it holds up under empirical scrutiny.

Communication and Information

Migrants are not only a source of pecuniary transfers; they also provide information to the home community, as discussed above. They are sources of information about jobs, ways of living, political developments and health resources. Again, while this information exchange is generally acknowledged, its empirical impact is difficult to assess. Separating the impact of the information itself from the financial contribution is an even more daunting task. Social networks are important in migration and origin-destination connections, but how these networks operate differentially across age, gender and ethnicity and how the actual information flows remains to be seen (Curran and Sagay 2001).

There is considerable work on the adaptation of migration in the destination, looking at everything from personal happiness, to income in the new setting or fertility behaviour. Conversely, there is only limited work that examines how migrants alter health and socioeconomic outcomes in the origin (Chattopadhyay et al. 2006, Lindstrom and Saucedo 2001). Much of this discussion is speculative or requires the analyst to go beyond the temporal structure of the data at hand. Again, HDSS can be quite helpful in providing more detailed ordering of departures and arrivals related to changes in household well-being.

The influence of migration need not be limited strictly to *circulation* of persons, material goods or remittances. The *settlement* of persons can have parallel effects. These can arise from the return migration of long-term out-migrants or the addition of entirely new members of the community through marriage, co-residence or new household formation. Migrants from other regions may bring new ideas, new knowledge and the like. As with some of the discussion above, such innovation and social network information flows are very plausible and much-discussed, but the empirical evidence to buttress the case is harder to find.

The Site Contributions to Migration and Livelihoods

Part II of this book takes up migration and livelihoods with contribution from three HDSS sites: Kanchanaburi, Thailand; Agincourt, South Africa; and Matlab, Bangladesh.

In Chapter 5 Sureeporn Punpuing and Philip Guest examine 'Migration and Agricultural Production in Kanchanaburi, Thailand'. They directly address the relationship between household migration and economic activity. In doing so, these authors provide the first example in this volume of how to use HDSS data to shed more light on temporal and socioeconomic changes in low-resource settings. Punpuing and Guest seek to show how migration away from agricultural origin households impacts labour force allocation of the remaining household members and the allocation of land to different forms of agriculture. They note that the evidence for the impact of migration on origin-households is mixed, even though the literature shows a general benefit of geographic mobility for the migrants themselves. This uncertainty underscores the value of empirical analysis with the kind of data provided by the Kanchanaburi HDSS (See Chapters 2, 3 and 4 for more details about the HDSS and the community context in Thailand). Their work focuses on outcomes for about 5,000 agricultural households.

Punpuing and Guest find some evidence from their analysis that out-migration results in a reduction in the amount of land used for agriculture. But their analysis also indicates that agricultural households with more resources, that is, more land under cultivation, are more likely to be able to fund one or more members to migrate. What may be more telling from their analysis and this is a result following directly from the ability to look at migration-household dynamics longitudinally, is that there is a quick adjustment after migration of a household member. Kanchanaburi agricultural households appear to adjust human resources to compensate for the loss of labour through migration.

In Chapter 6, 'The Dynamics of Migration and Socio-Economic Status in Rural South Africa, 2000–2007' Mark Collinson and several co-authors address the issue of remittances, clearly one of the thorniest in rural development and migration studies. They make use of over a decade of longitudinal data from the Agincourt migration site. Their work is particularly devoted to uncovering the relationship between migration and household socioeconomic status.

Following their multivariate analysis, Collinson and colleagues conclude that remittances from temporary migrants can, indeed, help lift families out of poverty. They also argue, as a side note and in contrast to some earlier literature, that larger Agincourt households are better off. They offer strong empirical regression results that suggest an above-poverty level of living is associated with one or more temporary migrants having circulated from the household to other communities and urban areas. Collinson et al. also argue that migrants to big cities are not necessarily the main generators of origin household well-being. It depends on the employment changes achieved through the migration. Of considerable consequence, perhaps, is their finding regarding gender and migration. Female temporary migrants are also important in predicting improved conditions for sending households. Thus, this pattern of temporary (or circular) migration is likely to serve as a net benefit for origin households.

In Chapter 7 Nurul Alam and Peter Kim Streatfield examine 'Parents' Migration and Children's Education in Matlab, Bangladesh'. They draw on data from one of the largest and longest-running HDSS systems in the world. They find a positive association with father's migration and child's educational attainment. More specifically, their regression analysis pointed to a higher expected rate of school enrolment and passing grades five, seven and ten for children of migrant fathers. It also seems to be the case that long-term and long-distance migration (probably linked to movement to oil-rich middle-east countries) exhibited a stronger relationship on education than more proximate migration. It is noteworthy that these effects appear in models that also control for fathers' and mothers' educational attainment. Migration of mothers was not related to children's education. Alam and Streatfield also found that the child's age and gender and household asset level were also related to enrolment.

Migration and Health

Migration is often implicated in the changes in health status of any population. We still know only a modest amount, however, about how migration and health are related. HDSS can be helpful in revealing some of these relationships and in pointing the way to health interventions, overall policy and promising avenues for further research. The connection between health status and migration is clearly bi-directional. In one direction, the act (or happenstance) of migration may influence health outcomes. In the other direction, a person's health may influence one's propensity to migrate or the destination one seeks. Even more, building on the labour market and livelihood discussion above, we might see migration as not an atomistic occurrence (one's own migration linked to one's own health), but, rather, see the migration–health relationship embedded in a family and community context.

Both directions can be tapped with HDSS data. This, in turn, suggests a potentially vital role for surveillance systems in linking migration to health. The repeated visits to HDSS households, the knowledge of the timing of departure and return and strategic inquiry about household well-being and behaviour could add much to this very policy-relevant discussion. Although some pitfalls exist in making proper inferences about such relationships, even with the benefit of longitudinal data collection, there appears much to be gained from exploiting HDDS. This takes migration decidedly out of the realm of a data management problem (nuisance!) and into the realm of being an intrinsic object of study to understand health outcomes in low-resource settings. Here we give some overview and illustration of the migration–health links that are pertinent to the substantive setting and the technical features of HDSS.

Disruption and Stress

Most immediate and self-evident, perhaps, is the stress that migration can induce for the individual migrant and his or her social network. Here we consider the Migration → Health direction of the relationship. Migration is one of the life-course's stressful events (White and Lindstrom 2005). In addition to stress as a health outcome in its own right, stress can be related to other health outcomes and demographic behaviours. This stress might be limited to the psychological, loss of social networks or it may perhaps have a physiological manifestation. Prolonged stress may reduce the capacity of one's immune system.

More research appears to have accumulated in demography for understanding the disruptive and stressful effects of migration on fertility. In the migration literature there has been considerable interest in the effect of migration on fertility, through the mechanism of disruption (Brockerhoff 1990, Goldstein et al. 1997). The notion of the psychic costs of migration (DaVanzo and Morrison 1981, Lindstrom 1996) captures the ideas that migration generates stress. Such disruptions and stress would apply to adult and child migrants (and in a secondary way to an entire family and community network), but there is limited definitive information for developing settings about the impact of relocation, per se. Disruption effects may work on networks related to health. For example, such a mechanism seems to be plausible in explaining the low immunization rates of children born to migrant women in Ethiopia (Kiros and White 2004). These women presumably had weaker social ties to provide health information and buttress health-seeking behaviour on behalf of their children.

In circumstances of industrialization and urbanization, especially when those changes are rapid, migrants may be especially vulnerable as they live in temporary housing or shantytowns, such as forms the study site for the Nairobi HDSS in this volume. Long standing patterns of labour migration, such as in South Africa, may have this character, as does more recent 'temporary' migration as seen in Asia. Temporary migrants in Vietnam for instance, seem to have inferior health outcomes (Nguyen and White 2007).

Exposure Regimes

Most directly related to migration, the geographic relocation of persons is likely to alter their exposure to disease regimes. Environmental conditions surrounding the risks of contracting malaria, pertinent to many of our surveillance sites, provide an example. Movement to an urban area, to a higher elevation or to a different climatic zone, can dramatically alter exposure. Similarly, movement to a large congested urban area – Bangkok, Mexico City, Dhaka, Accra, Dar es Salaam – can elevate exposure to airborne particulates and other risks.

Indeed, environmental health risks vary significantly across geographic territory and migration of persons across that same territory alters exposure. That is only the most obvious pathway to influence ultimate health outcomes.

Social processes associated with identifiable health outcomes can just as well be part of this process. The nutrition transition, with manifestation of such adverse conditions as obesity, is another example. While the nutrition transition is often considered society-wide, we can also think of urbanward migrants being exposed to a different lifestyle and food consumption pattern that results in changes in body mass index, hypertension or other health indicators. In addition, migrants with health knowledge specific to their origin may be less prepared to take protective measures in their new location. Thus, the social as well as the strictly epidemiological dimension of health is implicated in migration.

Health Seeking Behaviour

We now consider the arrow travelling in the opposite direction, that is, the health status (current or desired) might influence migratory behaviour. In health policy circles there is of course considerable concern about health seeking behaviour, especially the individual and household characteristics that seem to promote the effort to find and use community health services. We can readily extend this line of thinking to more distant movement: individuals and households may migrate to gain access to treatment for a particular adverse health episode or persons may move to place themselves in locations where the overall availability of services is higher. To be sure, this is one of the arguments used to explain the balance of rural-urban migration in many countries (Montgomery et al. 2003).

Individuals suffering from a chronic debilitating health condition (AIDS, cancer) may relocate to access care for their condition. Conversely, individuals, especially in resource-poor settings, may move to be nearer to kin and other caretakers. Such movement ironically may introduce a flow toward locations of *lower* health infrastructure, as perhaps the sick move back to seek care from rural kin. In fact there is some evidence of such kin-care based movement for terminally ill AIDS sufferers seen in the analysis of the Agincourt HDSS (Clark et al. 2007).

Vigour

In a less disease-specific case, migration is likely related to vigour, that is, overall health and perception of that health. We might expect healthy individuals to be more likely to undertake migration. For this reason, in addition to the life-cycle–labour-market motivation, younger individuals are more likely to undertake long-distance migration, an age pattern consistent with the migration profiles presented for several of the sites in this volume. This issue of variation in vigour (conversely, frailty) raised the spectre of population heterogeneity in confounding certain outcomes. Longitudinal data and analysis can considerably improve analyses of health in such circumstances.

The Site Contributions to Migration and Health

Part III of this book takes up migration and health with contributions from four HDSS sites: Nairobi, Kenya; Kisumu, Kenya; Manhiça, Mozambique and FilaBavi, Vietnam.

In Chapter 8 Adama Konseiga and his co-authors examine 'Assessing the effect of Mother's Migration on childhood mortality in the informal settlements of Nairobi'. In contrast to most other HDSS (including those represented in this volume), the Nairobi HDSS is urban. More than that, it is focused on large, geographically concentrated slum populations within the city. Slums have long been of concern both for being perceived as the source of various social and health problems and at the same time, the location of unfortunate victims of the ravages of impoverished urban living. Within this narrative, migration often takes on a prominent role. Still, the amount of research that can disentangle health outcomes and the circumstance of slum dwelling is quite limited. Konseiga et al. aim to improve on the existing record.

Konseiga and his co-authors begin by acknowledging that children born in the slums to migrant mothers exhibit a 60 percent higher risk of mortality than those born to non-migrant mothers. They then make use of event-history methods (a Cox proportional hazards model) to investigate how mother's recent migration status predicts survival rates for children born in the NHDSS, while also controlling for demographic and socioeconomic factors. They find quite strong results suggesting that mother's recent migration (within the two year analytical birth-survival window of 2003–2005) is a risk factor for childhood mortality. During the 2003–2005 period, children with mothers of recent arrival had three times higher mortality than births to long-term resident mothers. This statistical differential points to the importance of identifying potential mechanisms. One possibility is that newly resident mothers and their newborn children may not have yet adapted to the environment of the urban slum. Konseiga et al. question whether migrant mothers are sufficiently integrated into the new environment, perhaps unaware or uncertain about how to seek health care services for their children.

In Chapter 9 Kubaje Adazu and colleagues investigate 'Child Migration and Mortality in rural Nyanza Province: Evidence from the Kisumu Health and Demographic Surveillance System (KHDSS) in Western Kenya'. The authors of this chapter focus on the migration experience of the 2004 birth cohort of children, contrasting the experience of those who resided continuously in the DSA with those who entered the DSA sometime after birth. Like the authors of the previous chapter, they exploit the longitudinal nature of the Kisumu HDSS to conduct a hazard model of child's migration and health outcome. The analysis is motivated by the concern – akin to the migration-stress-disruption model mentioned above – that geographic relocation itself puts children at greater health risk. On the face of it the proposition seems plausible – potential stresses, new social networks, new exposures – and is of *prima facie* concern to public health professionals.

Adazu and colleague do not find support for the simple migration-adversity relationship. They found that children migrating from the urban to rural areas exhibited a 35 percent (and statistically significant) survival advantage, after controlling for household socio-economic status and mothers' and fathers' education and employment. Children migrating from other rural areas did not have a significantly different survival chances compared to non-migrant children. Adazu et al. hence argue that child migration is not necessarily negative and may even be positive if the risks incurred from migration are offset by the health endowments and migrant selectivity associated with the migration. This chapter makes note of the important aspects of heterogeneity and selectivity that can intrude in any such analysis, even one built on a base of longitudinal data. These results call attention to the important value of tracing individuals – young or adult – over time.

In Chapter 10, Ariel Nhacolo and collaborators examine 'Migration and Adult Mortality in rural Southern Mozambique: Evidence from the Demographic Surveillance System in Manhiça District'. They begin by noting that before 1999 in rural Mozambique returning migrants had a lower mortality than non-migrants; then a mortality cross-over took place and the mortality became higher for migrants than for non-migrants. This mortality cross-over and the underlying substantive links between migration, health and HIV/AIDS motivate their analysis. They work with more than a decade of HDSS data and also apply event-history techniques.

Mortality rates increased over the study period for returning migrants and non-migrants, but the increase was steeper for the returning migrants. Moreover, the authors find variation in the association between age and mortality. They find some evidence for the healthy migrant effect, but only at young ages. They also find a strong association between recent return migration and mortality, primarily due to HIV/AIDS. The authors take this to mirror the 'returning home to die' phenomenon seen for South Africa by Clark et al. (2007).

In Chapter 11 Ho Dang Phuc, Nguyen Xuan Thanh and Nguyen Thi Kim Chuc look at 'Migration and Under-Five Morbidity in Filabavi, Vietnam'. Phuc et al.'s regression analysis examines the predictors of sickness episodes by migration status, using data from the Filabavi HDSS. Phuc and colleagues find that the gender of the migrants parents' impacts differently on the burden of childhood morbidity. They confirm that socioeconomic traits (mother and father's education and occupation) significantly predict child sickness. However, even after controlling for these there is still predictive power of migration. A mother's out-migration is positively associated with under-five morbidity. This may be due, the authors suggest, to the disruption of child care caused by the migrations. There is a suggestion of slightly higher child morbidity for those with out-migrant fathers, but this relationship is not found to be statistically significant.

Conclusions

Demographic surveillance systems have much to offer the study of migration and its links to socioeconomic well-being and health outcomes. At the same time, there are some challenges, including intrinsic design difficulties, which present themselves to the analyst who would use HDSS data to understand social relations in these contexts. We now summarize some of the key aspects of understanding, managing and analysing migration in demographic surveillance systems. These comments may help set the stage for the reader who takes up the remaining cross-site chapters of this first part of the book and then moves on to the thematic presentations of site-specific results.

First with regard to the event of migration itself, it is crucial for those involved with HDSS – whether administrators, data managers or analysts – to recognize the universality and extent of human migration. As this volume demonstrates, even predominantly rural low-income communities in Africa and Asia experience significant in- and out-movement. Smaller geographic areas, such as the typical district-sized DSA are particularly subject to the potential influence of geographic mobility.

Even though migration is far less biologically linked than other basic sources of population change, there are strong demographic regularities with migration. There is a clear age profile, seen in Chapter 4 with data from the several contributing sites. This age regularity not only provides a window on the behavioural forces driving movement, but also reminds the HDSS manager and the scientist of the need to carefully define and track migration in the local system, so as to more accurately measure other population characteristics of interest: fertility, mortality, morbidity, nuptiality, school enrolment ratios and so on. The message is clear:

> Whether the HDSS analyst and management staff have an intrinsic interest in migration or not, there is overwhelming need to be attentive to the accurate measurement of movement in and out of the HDSS and to accurately record the timing of those moves and further, to accurately match electronic records of retuning persons to their original record upon return to the HDSS site.

Errors in population statistics are likely to be substantial if migration is ignored. Our discussion only begins to sketch the many health and social issues that are relevant to HDSS populations and are behaviourally linked to the migration of local residents. Stress and disruptions, monetary flows through remittances, the circulation of new information and the growth and decline of the overall population are strongly connected to migration. Migration is deeply implicated in social change.

These observations also point to some unrealized opportunities and challenges for HDSS and their advocates. They also point to directions for the future. To be sure, we think we have made the case for data management: tracking arrivals and departures and matching records of returning individuals. But there is more. Processes of social change and demographic selection influence health and socioeconomic outcomes in origin and destination. Studies of migration can help illuminate these processes. HDSS sites, with their comprehensive coverage and detailed temporal information are well-suited to such extensions:

> Demographic surveillance is well-positioned to carry out certain longitudinal studies of the determinants and consequences of migration for the migrants themselves and for their sending communities.

This is not a trivial extension of the work of HDSS. Some of this expansion – with longitudinal analysis – can be carried out within the existing data frame. For instance, site scientists can examine the differential health or socioeconomic outcomes of households who have sent a migrant to the city and HDSS sites will have proper timing and sequencing information to improve our current state of the knowledge. This exploitation of the longitudinal richness of HDSS does require sufficient resident data management and statistical skills on-site and across sites. The several site chapters presented in this volume begin to show what can be done in this arena. An even richer set of insights awaits as more temporally detailed data (including migrant follow-up) are analysed with the contemporary array of tools now at the disposal of the demographer and health policy analyst.

Further advancement of knowledge will require more investment. Particularly problematic for surveillance is the tracking of those who leave the DSA. The most useful longitudinal studies will require surveillance systems to follow migrants out of the DSA. Such an effort, whether designed to understand the adaptation of the migrants or the impact of remittances and communication with the origin community, require logistical extension beyond the standard HDSS management regime. Undoubtedly, this would require supplemental financing for such specialized studies:

> Advancing the range of applications of demographic surveillance systems to follow-up migrants who leave the demographic surveillance area (DSA) will require substantial new resources, but such an effort promises substantial returns to understanding the role of migration in social and health changes in developing settings.

As surveillance sites expand their mission from the narrowly biomedical to the more broadly socio-demographic, migration will feature more centrally in both the day–to-day work of the HDSS sites and more importantly, into the potential contribution to health and demographic science.

References

Adazu, K. 2002. *Migration and Child Mortality in a Rural Ghanaian Setting*. PhD Dissertation. Brown University.

Brettell, C.B. and Hollifield, J.F. (eds). 2000. *Migration Theory: Talking Across Disciplines*. New York: Routledge

Brockerhoff, M. 1990. Rural-to-urban migration and child survival in Senegal. *Demography*, 27(4), 601–16.

Cai, Q. 2003. Migrant remittances and family ties: a case study in China. *International Journal of Population Geography*, 9(6), 471–83.

Chattopadhyay, A., White, M.J. and Debpuur, C. 2006. Migrant fertility in Ghana: selection versus adaptation and disruption as causal mechanisms. *Population Studies (Cambridge)*, 60(2), 189–203.

Clark, S., Collinson, M., Kahn, K. and Tollman, S. 2007. Returning home to die: urban to rural migration and mortality in rural South Africa. *Scandinavian Journal of Public Health*, 35(Suppl. 69), 35–44.

Curran, S.R. and Saguy, A.C. 2001. Migration and cultural change: a role for gender and social networks. *Journal of International Women's Studies*, 2(3), 54–77.

DaVanzo, J.S. and Morrison, P.A. 1981. Return and other sequences of migration in the United States. *Demography*, 18(1), 85–101.

Goldstein, A., White, M. and Goldstein, S. 1997. Migration, fertility and state policy in Hubei province, China. *Demography*, 34(4), 481–91

Kiros, G.E. and White, M.J. 2004. Migration, community context and child immunization in Ethiopia. *Social Science and Medicine*, 59(12), 2603–16.

Lindstrom, D.P. 1996. Economic opportunity in Mexico and return migration from the United States. *Demography*, 33(3), 357–74.

Lindstrom, D.P. and Saucedo, S.G. 2001. Short- and long-term effects of US migration experience on Mexican women's fertility. *The Soc. F.*, 80, 1341.

Lucas, R.E.B. 1997. Internal migration in developing countries. *Handbook of Population and Family Economics.* Amsterdam: New Holland. 721–98.

Montgomery, M. et al. (eds) 2003. *Cities Transformed: Demographic Change and Its Implications in the Developing World.* Washington: National Academies Press.

Nguyen, L.T and White, M. 2007. Health status of temporary migrants in urban areas in Vietnam. *International Migration*, 45(4), 101–34.

Sana, M. and Massey, D.S. 2007. Family and migration in comparative perspective: reply to King. *Social Science Quarterly*, 88(3), 908–11.

Sankoh, O., Kahn, K., Mawageni, E., Ngom, P. and Nyarko, P. (eds) 2002. Population and health in developing countries Volume 1, Population health and survival at INDEPTH sites, in *INDEPTH Series Volume 1*. Ottawa, Canada: IDRC.

Taylor, J.E. 1999. The new economics of labour migration and the role of remittances in the migration process. *International Migration*, 37(1), 63–88.

VanWey, L.K. 2004. Altruistic and contractual remittances between male and female migrants and households in rural Thailand. *Demography*, 41(4), 739–56

VanWey, L.K., Tucker, C.M. and McConnell, E.D. 2005. Community organization, migration and remittances in Oaxaca. *Latin American Research Review*, 40(1), 83–107.

White, M.J. and Lindstrom, D.P. 2005. Internal migration. *Handbook of Population*. Dordrecht: Kluwer, 307–42.

White, M.J., Moreno, L. and Guo, S. 1995. The interrelationship of fertility and migration in Peru: a hazards model analysis. *International Migration Review*, 29(Summer), 492–514.

White, M.J., Muhidin, S., Andrzejewski, C., Tagoe, E., Reed, H. and Knight, R. 2008. Urbanization and fertility: an event-history analysis of coastal Ghana. *Demography*, 45(November), 803–16.

Chapter 2

Health and Demographic Surveillance Migration Methodology and Data: A Promise for Cross-Site Comparative Analyses

Kubaje Adazu[1]

Introduction

The fundamental characteristics of a Health and Demographic Surveillance System (HDSS) include a baseline census of a geographically defined population and repeated waves of data collection on changes in residential status and other demographic events such as births, deaths and marital status at relatively short intervals, often two to four times per year. Individuals who have resided in the surveillance area for a specified period prior to the launching of the HDSS are enumerated at the baseline census and automatically become members of the surveillance population. Subsequently, an individual could become a member of the population through birth or in-migration. Resident members may exit from the population through death or out-migration.

Currently the INDEPTH network has a membership of 37 surveillance sites in 19 different countries spread across sub-Saharan Africa, Asia, Central America and Oceania. As this volume indicates, these sites have a great potential to deepen our understanding of migration in the remote rural communities and deprived urban settlements where they are hosted through cross-site comparative analyses. Such comparisons do not only make it possible to determine whether the migration rates of a particular site are high or low but can also help to identify generalizeable patterns that could be used to inform theoretical propositions.

Most of these HDSS sites have their origins in biomedical monitoring and/or interventions. Furthermore, many were established with the notion that they would exist only for the duration of the interventions. As such, the conceptualization, capture and analysis of migration data are often underdeveloped in the sites. Therefore, when making these comparisons one must keep in mind that there are some methodological and procedural differences between sites. The most notable for this volume is the absence of a standardized definition of migration across the sites. The structures and migration data collection tools also vary across sites.

1 Kubaje Adazu passed away in January 2009. This chapter was in near-complete draft and has been updated by Michael White.

Nevertheless, demographic surveillance requires a record of people entering or leaving the study population and on that basis we can generate in- and out-migration rates and migration streams and counter streams from different sites and compare them (Chapter 4 in this volume).

This chapter first discusses comparative analyses and measures of migration. Then it presents an overview of how migration is defined and measured across the demographic surveillance sites. The chapter briefly describes the data collection methods and highlights the strengths and weaknesses of the data being collected by the HDSS sites. Sources of errors and potential biases that could arise from these errors are presented and discussed in the light of data quality. The chapter highlights the methodological challenges posed by cross-site comparative analyses and concludes with some suggestions on what can be done across the sites to synchronize migration measures and data collection tools in order to respond to these challenges.

Comparative Analyses and Measures of Migration

Although migration is recognized as a dominant force shaping family livelihoods and well-being throughout the developing world, too often our understanding of these interactions is limited to inferences from migration treatise with little empirical data (Tienda et al. 2006, De Bruijn et al. 2001) or to inferences from cross-sectional data, which provide us with snapshots in time, but give limited perspective on the families and communities linked by the migration. Furthermore, migration is very hard to classify, thus it is not easy to conduct cross-national comparative analyses because of differences in the methods of definition, measurement and data collection.

Migration is ideally measured as an event where every move made by an individual across a migration defining boundary is recorded at the time the move is made. At least in principle, this approach is commonly used in countries with national population registers. Some surveys such as the Demographic and Health Surveys (DHS) and national censuses have modified this approach somewhat by retrospectively asking for the number of moves in a given time period.

The most elaborate version of this is the retrospective migration or residence history, now gaining increasing use in a variety of data collection efforts. Migration can also be measured as a transition between two points in time by comparing the usual places of residence of an individual at these points in time. This approach is typically used by national censuses and sample surveys. Other measures of migration include place of last residence and duration of stay at current place of residence.

Migration is a repeatable and reversible event driven by social, cultural, economic, political and environmental factors. In addition, migration is two dimensional; it has both temporal and spatial dimensions. Hence measuring migration requires both geographical and time scales. Unlike the other repeatable event of fertility, migration has no natural biologically determined spacing: an individual may move again in any short (or long) space of time. Furthermore, while it is generally agreed that migration involves a permanent change in the usual place of residence of an individual, there is no consensus on how the temporal and spatial dimensions over which migration may be measured and the definition of usual place of residence varies from country to country.

Notwithstanding the measurement challenges, a number of scholars have attempted comparative analysis on the levels and patterns of migration (Rees and Kupiszewski 1999, Long 1991, Rogers et al. 1978), measures of migration such as migration distance, effectiveness and intensity (Bell et al. 2002, Stillwell et al. 2000, Long et al. 1988, Courgeau 1973), the demographic characteristics of migrants (Long 1991, Long et al. 1988, Rogers and Castro 1981, Rogers et al. 1978), the consequences of migration for the individual migrants and the places of origin and destination within countries (Nam et al. 1990) and the sources and quality of migration data (Nam et al. 1990). There is also a substantial body of literature on cross-national comparisons of the relationship between migration and fertility, migration and health (Garenne 2006, de Castro and Singer 2006) and migration and urbanization (Cerrutti and Bertoncello 2006, Champion 1989, Fielding 1982).

Overview of Migration Concepts, Definitions and Measurements at the HDSS Sites

Migration is defined as the movement of people in a given time period across a specified boundary, usually for the purpose of establishing a new residence (Haupt and Kane 2004). The movements could be permanent, temporary or circular. The migration defining boundary could be national borders or administrative boundaries within a country. Migratory movements that cross administrative (or sometimes other functionally defined) boundaries within a country are classified as internal migration whilst those that cross national borders are classified as international migration. The term migration interval is used to describe the time period over which the movements are observed. Depending on the migration event of interest the migration interval could be definite or indefinite. Definite intervals are used to measure fixed-time migration whilst indefinite intervals are generally used to measure lifetime migration events.

In all the HDSS sites the boundaries of the surveillance areas are used as the primary migration defining boundaries. In sites where the DSA encompasses parts of two or more administrative divisions, for example, Kisumu, the boundaries of the DSAs are drawn using the local landmarks. Hence identification of the migration defining boundary is not an issue in any HDSS setting. However, the surveillance areas vary significantly in land area and population size and these variations influence the probability that a move will be counted as a migration (across the boundaries of the DSA). The smaller the area, the more likely that any movement will cross the boundary. The HDSS also tracks changes in places of residence within the surveillance area and the boundaries of the residential units are used to define these types of localized movements.

By prospectively recording the changes in residential status, all the HDSS sites of the INDEPTH network measure migration as an event, but vary in the threshold of time required for an in- or out-move to be considered a migration. The minimum period of absence from the surveillance population to qualify as a migrant varies from one month at sites such as Kanchanaburi to six months at sites such as Matlab. At the Kenya sites (Kisumu and Nairobi) the threshold of time required for an in- or out-move to be considered a migration is four months, while at Manhiça and Filabavi it is three months. In the Agincourt site temporary migration is recorded as an individual residence status which is repeatedly updated, but for comparison with other sites the permanent and temporary migration streams are merged into one migration rate with a three month threshold period for migration (see Chapter 4 for details).

The Outlines of a HDSS

Health and Demographic Surveillance Systems (HDSS) are now well-established features of the health and demographic data collection scene. Therefore, for present purposes, we only review some key features of HDSS, especially with relevance to our topic of migration in this book. More detail about the design and administrative operation of surveillance systems are available elsewhere (Sankoh et al. 2002). The INDEPTH Network is the central coordinating body for the sites participating in this volume and for many more beyond.

The key features of HDSS are:

- A territorially defined population of interest, often about the scale of an administrative district.
- Demographic accounting for 100 percent of the population under demographic surveillance (census).
- Repeated waves of data collection, often three to four per year to produce panel, that is, longitudinal, data.
- Availability of community-level information, whether an aggregation of individual data or separately collected community-based information, such as the presence of electricity, proximity to a school or paved road.

These features form the scaffolding of the HDSS and they constitute the structure on which health studies (both observational and experimental design) and demographic studies are built. These structural features and some of the assumptions on which they were originally based as the HDSS were being designed, also make for particularly challenging and promising aspects of incorporating migration into the analytical frame of surveillance.

These features of a HDSS offer many strengths. They usually make for a wealth of observations: in this volume the range of number of persons under HDSS observation ranges from 43,000 (Kanchanaburi) to 212,000 (Matlab). The first INDEPTH volume provides more detail on all the sites and provides additional discussion of the features of surveillance (Sankoh et al. 2002).

HDSS Migration Data and Methods

This section reviews the migration data collection procedures in the HDSS settings and outlines the types of migration indicators possible from HDSS data. The key features of a demographic surveillance system include a baseline census of a geographically defined population and mechanisms for continuously monitoring births, deaths and changes in residential status. The monitoring mechanisms include household registers that are updated regularly, well defined rules for assigning household membership and residency and unique permanent identification numbers for residential units, households and individual members of the surveillance system. In every round of data collection field assistants move from door to door and update the registers. The frequency of updating the registers varies from once a year at sites such as Agincourt in South Africa and Kanchanaburi in Thailand to four times a year at sites such as Filabavi in Vietnam (Table 2.1). The greater the frequency of surveillance, the greater the number of movements likely to be observed in any fixed span of time. Shorter minimum duration for the observed movement would also tend to increase the number of movements counted as migrations.

Changes in places of residence within the surveillance areas are also tracked so as to avoid double enumeration of individuals who move from one residential unit to another within the surveillance area. In addition to recording the migration and local mobility events for individuals who have either exited from or moved into the DSA, the date of departure or arrival are also recorded. The sex and date of birth are collected for individuals coming into the population (in-migrants) for the first time. In some sites the reasons for moving and characteristics of the origin/destination of the move are obtained from the migrants or their household or from neighbouring residents in cases where an entire household has out-migrated.

With these detailed queries it is possible to accurately determine who left or entered a particular residential unit within the surveillance area and when. It is also feasible to determine who left and came back and of course, their duration away. The HDSS sites are therefore uniquely positioned to provide accurate data on in-migration, out-migration, gross-migration (sum of in-migration and out-migration), net-migration (the difference of in-migration and out-migration) and temporary or circular migration. Sites can also generate migration rates stratified by age and sex using either person-time or the mid-year population as the denominator. Sites that have origin-destination information can generate estimates of migration streams (the total number of moves made during a given migration interval that have a

Table 2.1 Site frequency of updates and definition of migration

Sites	Surveillance frequency (per year)	Minimum duration for movement to count as a migration
Kanchanaburi	1	1 month
Agincourt	1	3 months
Matlab	2	6 months
Nairobi	3	4 months
Kisumu	3	4 months
Manhiça	2	3 months
Filabavi	4	3 months

common area of origin and a common area of destination) and counter-streams. From the origin-destination information, the rural based sites could further classify the migration streams into rural-rural and rural-urban and the urban sites could generate rural-urban and urban-urban streams. Sites that collect socioeconomic data, such as education, household possessions and housing structure in addition to the demographic data can further stratify the migration rates by level of education and household socioeconomic status at the time of migration to shed more light on how migration influences household socioeconomic status and vice versa.

Some sites are also in a position to generate estimates of lifetime migration. The conventional approach of estimating lifetime migration is to compare the place of birth and current place of residence and all individuals whose places of birth are different from their current places of residence are classified as lifetime migrants. Though useful, this approach, excludes all people who moved away and subsequently returned to their birthplaces prior to the survey or census date. In a HDSS setting it is easy to determine all individuals who left the DSA after the baseline enumeration survey and came back. Individuals who have relocated within the surveillance area can also be easily identified. Because the HDSS tracks all moves it also possible to determine how many times an individual has moved within and across the boundaries of the DSA in any given time. The surveillance model therefore measures and represents all migration episodes (cumulative number of migration episodes) and duration of exposure in different residential environments. Thus, HDSS sites that collected information on place of birth or date of arrival at the DSA during the baseline enumeration are well positioned to provide better estimates of lifetime migrants in the surveillance populations by combining the lifetime migrants derived from the conventional method with the returned migrants identified through the continuous registration system.

1. Migration and Population at Risk

As noted in Chapter 1, for some health scientists, human migration is a nuisance: People originally enrolled in a study disappear or new people arrive and need to be included or consciously excluded. In contrast, all population scientists recognize the potential for migration to alter the population at risk and in turn, influence key demographic rates such as total fertility or child mortality. As persons enter and leave the HDSS they obviously change the population at risk. Ideally any demographic rate – Total Fertility Rate (TFR), life expectancy or infant mortality – should be expressed as an appropriate function of the population at risk.

In closed populations, by definition, migration is, not a problem. In large populations, say a national population, the assumption of a closed population may not introduce any serious error. But in a district or regional population, as with a population included within a HDSS, the assumption of a closed population can lead to gross errors. For instance several sites have 20–24-year-old female out-migration rates in the range from 10–30 per hundred (person years). Retention of such individuals in the denominator when they have departed the community (and their childbearing would not be registered in the HDSS) can impart a sharp bias to the HDSS Age Specific Fertility Rate (ASFR) and the derived TFR.

Even more troublesome is the fact that the prime ages of childbearing are also coincident with some of the highest migratory years, so the potential for error is severe. Beyond this, errors in accounting for the whereabouts of such women influence adult mortality (their own risk) and infant and child mortality, given the whereabouts and experience of their own births and youngsters. In settings with high infant mortality rates (50–100 per 1,000) in the kind of rural low-income communities where surveillance is often undertaken – the departure and entry of women to the DSA can certainly complicate the effort to derive an infant mortality rate that meets the demographer's specification of the population at risk.

2. Migration and Selectivity

There is a further complication. The act of migration may be related to the very process being measured or deemed worthy of analysis. Heterogeneity and selection processes are now very well recognized confounding factors in health, social and demographic studies. It is probably the case (although confirming statistical analysis of this would be helpful) that as the use of HDSS data drift from narrowly tailored biomedical monitoring and intervention to more behavioural and social analyses, the potential confounding effect of migration and selectivity looms larger.

Consider the contrast of a vaccine trial and an educational intervention, both for children. Some children will likely move away during the course of observation in a vaccine trial. Analysts will appropriately treat such observations as censored and the censoring mechanism is likely to be unrelated to the biomedical process. As such, inferences made about the effect of the vaccine are likely to be robust to the loss of child participants (this is not guaranteed, however). By contrast, such procedures and assumptions may be less warranted in the case of schooling. It may well be the case that a child's migration away from the surveillance site may well be related to the family's socioeconomic circumstances and motivation for achievement. In one possible scenario, the most motivated and promising students may move out of the district (with their elders) in order to seek economic opportunity and education elsewhere. Analyses performed only on the remaining HDSS children might be problematic. More sophisticated examples of confounding effects of heterogeneity and selection in prospective studies can be found in the methodological literature. Suffice to say that outcomes more related to social process may be more suspect to the influence of migration. At a minimum, HDSS-based studies should consider models which predict the probability of loss-to-follow-up (LTFU) as a function of baseline characteristics. This suggests the value both of tracking migration well and incorporating its prevalence into analyses.

Strengths of the HDSS Migration Data

Health and Demographic Surveillance Systems were born out of the need for accurate and timely longitudinal information on populations living in remote rural areas or deprived urban settings where the risk of illness and death are very high but where there are no functional vital registration systems. Hence, the data generated by these systems have several strengths over the types of migration data collected in censuses and surveys. For, instance, duration of exposure and person-time can easily be computed from the HDSS data. Another strength of the continuous surveillance method is that it generates longitudinal individual-level and household-level data that can be linked to shed more light on the association between migration behaviour and household composition. More importantly, the data generated by the continuous surveillance method are suitable for time series and event history analyses at the individual, household and community levels.

With the surveillance method, all migratory moves made by a resident member in a given time period are recorded. Since the movements are recorded shortly after they have taken place, this method generates more accurate counts of both movers and moves, making it possible to easily identify frequent movers from infrequent movers. In contrast, censuses and sample surveys typically use migration measures that compare present and past residence and thus estimate the number of movers rather than the number of moves. For short intervals, movers and moves are likely not to differ significantly since not many people are likely to move more than once in the short interval. However, the same cannot be said for long intervals. Although some sample surveys and national censuses have questions that solicit information on the number of moves, the responses to these questions are subject to recall bias; hence, the estimates of the number of moves derived from this approach are likely not to be as accurate as the counts from the surveillance method.

Surveillance systems, by their very design, are particularly advantageous for *prospective* studies. It is important to realize that this longitudinal data collection plan benefits observational studies as well as intervention studies, although most attention seems to have gone into the latter. Thus, when data management is accurate and analysis careful, HDSS can give good indications of changing birth and death rates in a region. Ready extension can be made to tracking child health (observationally), education, socioeconomic well-being and the like. All this is on top of the surveillance advantages for health indicators or the study of response to a new health program or potential drug.

Like other prospective data collection models, the HDSS method is less susceptible to recall bias. In addition to the accurate timing of migration, the HDSS also accurately records the timing of other socio-demographic events that could affect or be affected by migration. It is therefore more feasible with surveillance data to find out the temporal sequence of migration events and any changes in household socioeconomic status resulting from the migration. The analyst can correctly order the series of life events; temporal sequencing of events is critical for accessing the determinants and consequences of migration, such as fertility, employment and health status. The accurate timing of migration also makes it possible to investigate seasonal patterns and seasonal correlates of migration.

Building on this prospective design and exploiting the skills of fieldworkers and the repeated participation of local residents, HDSS have been well-suited as platforms for experimental interventions. The central database is a ready source for a sample, which can be done with geographic specificity, randomization across villages or households and selecting an eligible universe (or strata within). For instance, many HDSS administrations and their data management offices stand 'at-the-ready' to draw a sample of women 18–24 years of age or children 9–13 years of age. HDSS samples could be drawn readily on more sophisticated criteria, such as those who experienced the death of a family member in that last year or who had elderly kin in residence.

Limitations of the HDSS Data

The surveillance systems have successfully collected high quality longitudinal data on migration and local mobility. Nonetheless, cost constraints have limited their coverage to relatively small geographical areas. Generalization beyond the population under surveillance can be problematic. Typically, HDSS are situated in rural areas and the site selection criterion is not always clear. Inferences to urban populations or even to the broader rural population may be suspect. In sum HDSS often – willingly or not – trade the gain in accuracy of repeated measurement of a known population for the loss in making inferences about a broader population.

So far, no INDEPTH site is tracking migrants beyond the boundaries of the DSA. Thus out-migrations are treated as LTFU in all the INDEPTH sites. Such a decision might make sense in short short-term interventions. For clinical trials that require study participants to be observed for longer periods ignoring migration or treating it as random censoring could bias the results. However, to retain individuals who have moved beyond the boundaries of the surveillance areas in clinical trials will require the surveillance systems to follow migrants out of the DSAs. Tracking migrants to destinations outside of the surveillance areas will no doubt have cost implications for such studies.

In addition to the challenges associated with LTFU, there are also inherent design problems with the HDSS data. The demographic surveillance systems were set up in most sites to provide sampling frames for public health interventions. As a result, data are collected on a limited number of socio-demographic variables. For example, migration origin/destination information is not being collected by some sites. Sparse information on migrant destinations or origins could pose a challenge to the generation of migration streams and counter streams. Another limitation is that the data are stored in complex relational databases and therefore require considerable time, effort and skill to extract analytical data.

Sources of Errors and Data Quality

Errors and biases are not uncommon to population data sources, especially those involved in the collection and compilation of large volumes of data. The demographic surveillance system is an intensive longitudinal data collection and processing machine. Like any large population data source, the HDSS is susceptible to errors arising from incomplete registration. In theory, coverage errors are less of a problem in an HDSS setting because the data collectors move from door to door to update the registers. But despite the effort to try to reach every household during the round of data collection, there are still instances where some household registers are not updated because the data collectors are unable to find (knowledgeable) persons from the households to interview. This problem is quite frequent among single person households, especially in peri-urban and urban settings. Members of single-person households are invariably the sole breadwinners and as a result they are often at work during the day and therefore could not be reached for interviews.

Because of the emphasis on a threshold duration of residence to qualify as a resident member, some frequent movers who move in and out at very short intervals may not be included in some HDSS populations, though they might have spent the same length of time in the population as de facto resident members. For example, consider two individuals who arrive in a site like Kisumu on the 1 January. One leaves the area in March and returns in April and leaves again in June while the other remains in the DSA from January through April. Cumulatively, these two individuals spent four months in the area in the period January to May, but the individual who left in March and came back in April would not be registered as a resident because he/she did not stay in the area for four consecutive months. The second individual would be registered as a resident by April, because he/she met the requirement to become a resident. Returning migrants who die shortly after arriving in the DSA and therefore fail to meet the residency requirement also would not be included in the population register. Hence, their migration and mortality events are not observed. Incomplete registration of new births is another source of coverage errors. In places where children are not named before or soon after birth, there is usually a delay in the registration of babies and as a result some of them end up not being registered at all. To overcome this problem some sites have expanded the household registration system to include tracking of pregnancies. The same procedures could be implemented for migrations to maintain a complete count of migrants.

As with other demographic data collection efforts, HDSS must contend with misstatement of age, non-response or incomplete responses and respondents intentionally or unintentionally providing inaccurate or false information, especially on the places of origin and destination. Wrong entries by data entry clerks during data entry and by interviewers during interviews and transcription and deliberate fabrication of data by aberrant data collectors are also possible sources of errors. Fabrications, transcription errors and wrong entries are often identified and corrected in subsequent rounds of data collection; hence, these errors do not pose a long-term threat to the integrity of the data. Coverage and non-response errors on the other hand are very difficult to deal with since in most cases the individuals can not be traced.

Accurate and reliable data are critical in migration analyses, more so in cross-site comparative analyses as those presented in this volume. Experience has however shown that in a longitudinal surveillance data collection setting, small errors when not detected and corrected multiply in time. Thus, all sites have quality control procedures such as repeat interviews and spot checks to ensure that data collectors visit the household and conduct required interviews. Data entry errors are minimized through built-in consistency and validation checks in the data entry screens and database structure. Some sites conduct double data entry in addition to the built-in checks. These built-in checks coupled with the inherent prospective nature of the surveillance model have resulted in improving the quality of data being generated by the INDEPTH sites. For instance, the three to four visits to the households per year have helped in reducing the problem of recall bias to a minimum.

Methodological Challenges

The most notable challenge to cross-site comparisons of the HDSS data is the lack of a standardized definition of migration across the sites. There is a wide variation in the minimum duration of absence from or residence in the DSA used to determine who qualifies as a migrant, ranging from one to six months (see Table 2.1). In addition to the problem of definition, sites independently design their data structures and migration data collection tools. As a result, only a few sites have identical data structures and data collection tools. The lack of standardized data structures makes data pooling difficult.

Beside the challenges arising from the variable definitions of migration and the lack of standardized data structures, there also is variation across the sites in who is at risk for experiencing a migration event, potentially resulting in individuals of similar circumstances being considered migrants in one site and non-migrants in another. This problem is partly due to how sites handle individuals who commute periodically between two homes and institutionalized populations, such as students in boarding schools. For example, some sites consider the homes of the parents of students in boarding schools as the usual place of residence and treat them as resident members of the parents' households who are temporarily absent from home, while other sites consider the colleges as the usual place of residence. At the sites where students of boarding schools are considered members of their parents' households, no in-migration events are recorded when the students eventually complete college and return home, since they were considered as 'residents' for all the time they were away in school. On the contrary, students at the sites where the homes and colleges are both considered as the usual places of residence, out-migration events are recorded for children who leave for boarding schools outside the surveillance area. When these children complete college and return home they are treated as return migrants and in-migration events are recorded for them.

The longitudinal nature of the surveillance data necessitates the use of complex relational databases to store the information. Extracting analytical datasets from such complex databases requires good computer programming and data management skills. As a result some sites are having difficulty exploiting the longitudinal advantages of their data.

The Way Forward to Cross-Site Comparison

Together, these sites have a potential to contribute to knowledge through cross-site comparative analyses. Cross-site comparisons are particularly useful because migration rates computed for one site become more meaningful when compared and contrasted with rates from other sites. The comparison across sites holds promise for providing more insights on patterns of migration and how these patterns affect and are affected by socioeconomic factors in the diverse settings of the network. In addition, the INDEPTH sites are uniquely positioned to extend the frontier of migration studies through meta-analyses of primary data from diverse settings and the provision of empirical data for testing hypotheses and fitting of migration models. However, for the full potential of these data to be realized the Migration and Urbanization Working Group (MUWG) of the INDEPTH network must find ways to address some of the methodological challenges highlighted above.

For reliable cross-site comparisons to be made, it is critical to have comparable measures of both the event of migration itself and the population at risk of experiencing migration. Otherwise it would be difficult to determine whether differences in migration patterns are due to real differences in migrant behaviour in the populations under observation or differences in how migration is measured across sites. To ensure that individuals of identical circumstances are treated exactly the same across the network sites must agree on how to handle individuals who have more than one (usual) place of residence and commute between the two places. A strict application of the cooking pot definition of household and a standardized minimum duration of absence from or residence in the DSA could minimize this problem.

Overcoming the challenges posed by differences in definitions and data structures and collection tools requires the standardization of migration definitions and harmonization of migration data structures and collection tools. In doing so, it is important to bear in mind the resource capacity and research agenda of sites; some sites are more endowed with resources than others. Therefore, well-endowed sites may want to collect more data than can be afforded by sites with limited resources. The migration data collection tool therefore could be broken into two sections. The first section could comprise standardized questions to be used to collect all the relevant information needed for cross-site comparisons, while a second part could be left open for sites to decide what additional information that they need, but which is not critical for cross-site analyses.

The lack of computing support is a challenge of major concern and given the competition for computer programmers from other sectors and the disadvantaged position of many HDSS sites, attracting and retaining seasoned computer programmers may not be easy. A sustainable approach to the difficulty of generating analytical datasets from the complex relational databases is being examined at INDEPTH (for example, in the INDEPTH Data System Initiative). The aim is to design universal and user-friendly interfaces for the management of longitudinal data that sites can download and use to extract analytical data sets from the main databases. Such interfaces with the capacity for end users to retrieve raw data from the main database in a flexible range of outputs will enable researchers to independently generate their analytical data sets.

While the suggested revisions to conceptualization, measurement and data process require additional resources and coordination across sites, these investments are likely to bring significant returns. Those interested in migration will surely benefit, but also the wider group of scholars, scientists, policy-makers and applied professionals concerned about the well-being of persons living in contemporary developing settings.

References

Bell, M., Blake, M., Boyle, P., Duke-Williams, O., Rees, P., Stillwell, J. and Hugo, G. 2002. Cross-national comparison of internal migration: issues and measure. *Journal of the Royal Statistical Society*, Series A, 165(3), 435–67.

Champion, A.G. 1989. *Counterurbanization, the Changing Pace and Nature of Population Deconcentration*. London: Edward Arnold.

Cerrutti, M. and Bertoncello, R. 2006. Urbanization and internal migration patterns in Latin America, in *Africa on the Move: Africa Migration and Urbanization in Comparative Perspective*, edited by M. Tienda, S. Findley, S. Tollman and E. Preston-Whyte. South Africa: Wits University Press.

Courgeau, D. 1973. Migration et decoupages du territoire. *Population*, 28(3), 511–37.

De Castro, M.C. and Singer, B. 2006. Migration urbanization and malaria: a comparative analysis of Dar es Saaam, Tanzania and Machadinho, Rondonia, Brazil, in *Africa on the Move: Africa Migration and Urbanization in Comparative Perspective*, edited by M. Tienda, S. Findley, S. Tollman and E. Preston-Whyte E. South Africa: Wits University Press.

De Bruijn, M., van Dijk, R. and Foeken, D. 2001. *Mobile Africa: Changing Patterns of Movement in Africa and Beyond*. Leiden: Brill Press.

Fielding, A.J. 1982. *Counterurbanisation in Western Europe*. Progress in Planning 17, London: Pergammon Press, 1–52.

Garenne, M. 2006. Migration, urbanization and child health in Africa: a global perspective, in *Africa on the Move: Africa Migration and Urbanization in Comparative Perspective*, edited by M. Tienda, S. Findley, S. Tollman and E. Preston-Whyte. South Africa: Wits University Press.

Haupt, A. and Kane, T.T. 2004. *Population Reference Bureau's Population Handbook*. (Fifth edition). Washington DC: Population Reference Bureau.

Long, L.H. 1991. Residential mobility differences among developed countries. *International Regional Science Review*, 14(2), 133–47.

Long, L.H., Tucker, C.J. and Urton, W.L. 1988. Migration distances, an international comparison. *Demography*, 25(4), 633–40.

Nam, C.B., Serow, W.J. and Sly, D. 1990. *International Handbook on Internal Migration*. New York: Greenwood Press.

Rees, P. and Kupiszewski, M. 1999. *Internal Migration and Regional Population Dynamics in Europe: A Synthesis*. Population Studies No. 32. Strasbourg: Council of Europe Publishing.

Rogers, A. and Castro, L.J. 1981. *Model Migration Schedules*. Research Report RR-81-30. Laxenburg, Austria: International Institute for Applied Systems Analysis.

Rogers, A., Racquillet, R. and Castro, L.J. 1978. Model migration schedules and their applications. *Environment and Planning*, 10(5), 475–502.

Sankoh, O., Kahn, K., Mawageni, E., Ngom, P. and Nyarko, P. 2002. Population and health in developing countries: Population health and survival at INDEPTH sites, in *INDEPTH Series Volume 1*. Ottawa, Canada: IDRC.

Stillwell, J., Bell, M., Blake, M., Duke-Williams, O. and Rees, P. 2000. A comparison of net migration flows and migration effectiveness in Australia and the United Kingdom, 1976–1996: Part 1, total migration patterns. *Journal of Population Research*, 17(1), 17–38.

Tienda, M., Findley, S., Tollman, S. and Preston-Whyte E. 2006. *Africa on the Move: Africa Migration and Urbanization in Comparative Perspective*. South Africa: Wits University Press.

Chapter 3

Dynamics of Space and Time: Community Context of Migration, Livelihoods and Health in the INDEPTH Sites

Sally E. Findley

Introduction

Where people live fundamentally affects their choices. Some communities are rich in opportunities and enable families not only to sustain themselves but to attain some of their dreams for a better life for themselves and their children, while others have less to offer and achieving a full, healthy life is more challenging (Entwistle 2007). Opportunities are not structured homogeneously within a community and differential access to opportunities is reflected in disparities in livelihoods (Payne and Lipton 1994, Milazi 1998, Scoones 1998, Ellis 1999). Migration is one of the primary mechanisms available to people facing an opportunity deficit, whether for their very survival or for the attainment of long-term goals. Migrants move to communities where they have more opportunities and less risk of deficits and downturns (Stark and Yitzhaki 1988, Findley and Sow 1998, Scoones 1998, White and Lindstrom 2005). While economic opportunities remain key to most migrations, other opportunity structures (for example social, educational, marital, security, health) also influence migration. When fleeing conflict, migration is dictated by the need for peace and security and less, at least initially, by the pursuit of better job opportunities (Findley 2001).

Whether voluntary or involuntary, migration involves changing one's residence from one place, the origin or 'home' to another, the destination or 'host' community. While some migrants completely shift their lives from one place to another, with few links between the two after they move, such as in the case of the refugee whose home is destroyed by war, many more migrants retain linkages of some sort between their new and former homes (Settles 2001). These links are maintained through the social networks and connections that migrants keep with those they left behind (Cliggett 2000, Gidwani and Sivaramakrishnan 2003, Kazemipur 2006, Hemer 2008). What the migrants tell and bring to their home communities informs subsequent migrants, who may follow in their steps based on their understanding of the returns from migration (Lin 1999, Shah and Menon 1999, Cliggett 2000, Kazemipur 2006, Amuedo-Dorantes and Mundra 2007). These social links also can be the means by which migration exerts an influence on the home community's economic structures long after the migrant has left (Massey 1990, Massey et al. 1998). Migrants can transform not only their families' lives through remittances, but they also can stimulate social and economic change in their home communities through investments in physical and social capital (Castles 2002, Sana and Massey 2005).

Thus, the context of migration includes place-related opportunities and the web of social connections linking migrants to opportunities at different places. In most instances, very little is known about how these intersect over time, but with the demographic surveillance sites we have a rare opportunity to consider how community context can shape movement by household members into and out of well-defined places, the INDEPTH network sites. In this chapter we will introduce the community context of the sites, highlighting the ways that their opportunity spaces differ and how this is likely to shape the migration patterns and migration–health interactions at each site.

Community Contexts of Migration and Health

Community operates through layers of influence and the influences on family livelihoods, health and migration are simultaneous, interactive and often linked. The integrated model of sustainable livelihoods, migration and health (Figure 3.1) is based on the rural livelihoods model of Scoones (1998), adapted to include health as a dimension of livelihoods. In this model, migration becomes a critical element of the strategy both to establish a sustainable livelihood and to support the household's health and well-being, as indicated by the bi-directional arrows between migration (shown in italics) and household livelihoods and then through migration to generate well-being, namely sustainable livelihoods and household dynamism. The levels of influence are the macro-context, community resources and household resources. Previous migration patterns influence subsequent migrations through their contextual influences of channelling migrants to specific destinations where others learn about opportunities (information), can connect to housing or jobs and can receive assistance or support.

The broadest and most encompassing influence is at the macro-context. Macro-contexts are generally regional or national, such as macro-economics (including terms of trade for agricultural and trade commodities, employment levels, indebtedness), political situation and conflict, climate variability, cultural diversity and national urban structure. The macro-context highly influences the range of migration options and pressures and conditions of relative deprivation set the stage for successive labour migrations (Stark and Yitzhaki 1988, Findley and Traore 1995). Nationwide economic difficulties, political instability and conflict press more individuals towards emigration away from the country or at least to more stable regions (Adekanye 1998). Climate variability and drought increase migration, especially circular migration (Ezra 2000, Findley and Sow 1998). Similarly, the nature of the urban hierarchy shapes where migrants go when they seek education or jobs in the city, as is seen below when considering different migration patterns from the sites. Future migration patterns are influenced by the distributions of previous cohorts of migrants and the migration channels they have created through years of use (Gurak and Caces 1992, Shah and Menon 1999, Castles 2002). For the INDEPTH network sites the macro-contexts are the national level for most macro-context variables, although climate and conflict may operate at the regional level.

Migration patterns are more directly influenced by the local community resources. Community for the INDEPTH sites is generally the geo-political unit, the village, township, neighbourhood or city, where the household is located and which encompasses most of the daily activities for most of the residents. The community resources are the local resources that shape livelihood and health options for households. Economic options are shaped by natural resources (land, water, natural resources), the technologies available for production and available economic and financial resources (Bilsborrow 1991, Cordell et al. 1996, Milazi 1998). Together these three sets of resources shape work and basic livelihood options, both the diversity and their productivity. A key element influencing household migration patterns is food availability and distribution, because if there are food deficiencies this will also engender migration (Payne and Lipton 1994, Hugo 1998, Findley and Sow 1998, Konsiega 2007). Community educational and social structures shape access to the productive resources, influencing the degree of equity and mobility potential open to families (Ellis 1999). Different types of community infrastructure (transportation routes and links to and through the region, sanitation and water availability, communications and media) including such things as bus access, permanence of roads and cell-phone towers further influence the development of economic opportunities and the access by segments within the community to economic, social and health resources. Migration networks linking the community with other communities, in or out of the nation, also exert a powerful influence on subsequent migration choices (Sana and Massey 2005). Many of these community resources also influence the household's health strategies, which are also influenced by the nature of the community health infrastructure and the presence of any vector control programs (Findley 1991).

The household's livelihood and health strategies are shaped by their own resources and how they are able to tap into community resources, as constrained or supported by the macro-context (regional or national). Critical to the diversification of household strategies are access to capital (loans, savings, business links) and their potential to link to opportunities elsewhere, whether through migration or via communications media or transportation links (Massey 1990, Hugo 1998, Hampshire and Randall 1999). Households with educated or skilled members will be able to compete in more diverse labour markets and in the absence of local opportunities will migrate, particularly if they have connections that can lead to employment at the destination (Cliggett 2000, Ezra 2000). Thus, households which have migrant linkages will be in a favoured position in competing for migrant opportunities (Sana and Massey 2005). Key elements in maintaining food and health security are access to food, social support networks, higher status for women and basic household water and sanitation systems (for additional discussion of the contextual determinants of health see Findley 1991).

Households use their own resources to develop their livelihood and health strategies, which incorporate access to community resources and resources outside the community, via migration (actual or remittances). If there is sufficient potential for agricultural intensification or expansion to additional lands, then that may be the option chosen, as it is for families which do not rely on migration to meet their needs (Scoones 1998). Those households diversifying into non-agricultural work may do so through local opportunities or, as is often the case, through migration, permanent or circular, to labour markets where migrants may find work and remit back to their families (Hugo 1998, Sana and Massey 2005). Migration also may be used to access education or social network resources, which can be used by the family to diversify or strengthen their livelihoods (Cliggett 2000). In Mali, for example, migrants brought back not only money but training in how to develop and operate irrigated fields, enabling diversification of the economy (Findley and Sow 1998).

Simultaneously, households are using their own and community resources to protect their family from poor health and disease. The health livelihoods integrate efforts to maintain adequate food, clean water and sanitation for the family, use of primary health care services as needed and regularly for preventive services, including the purchase of key health 'inputs' and participation in health education sessions. As has been noted elsewhere, women with higher social status and/or basic education are better able to implement the effective health strategies. Migration may also affect the household health through in- or out-migration of those with selective health status (better or poor health) or by migration to seek or obtain better health care.

Households aim for sustainability through varying combinations of livelihood, health and migration strategies. Sustainable livelihoods afford adequate work, food and well-being for families. Household dynamism gives households the flexibility needed to grow and change so they can respond to changing circumstances, much like an insurance policy. Household health is the sustained good health of all household members, the ability to ward off serious disease and reduce disability, resulting in the enhanced ability of its members to have a long and active working life.

Contexts of Migration, Livelihoods and Health: Perspectives from the INDEPTH Sites

Migration, livelihoods and health have been of interest at several INDEPTH sites and the rest of this chapter applies this integrated model of sustainable livelihoods, health and migration to these sites.

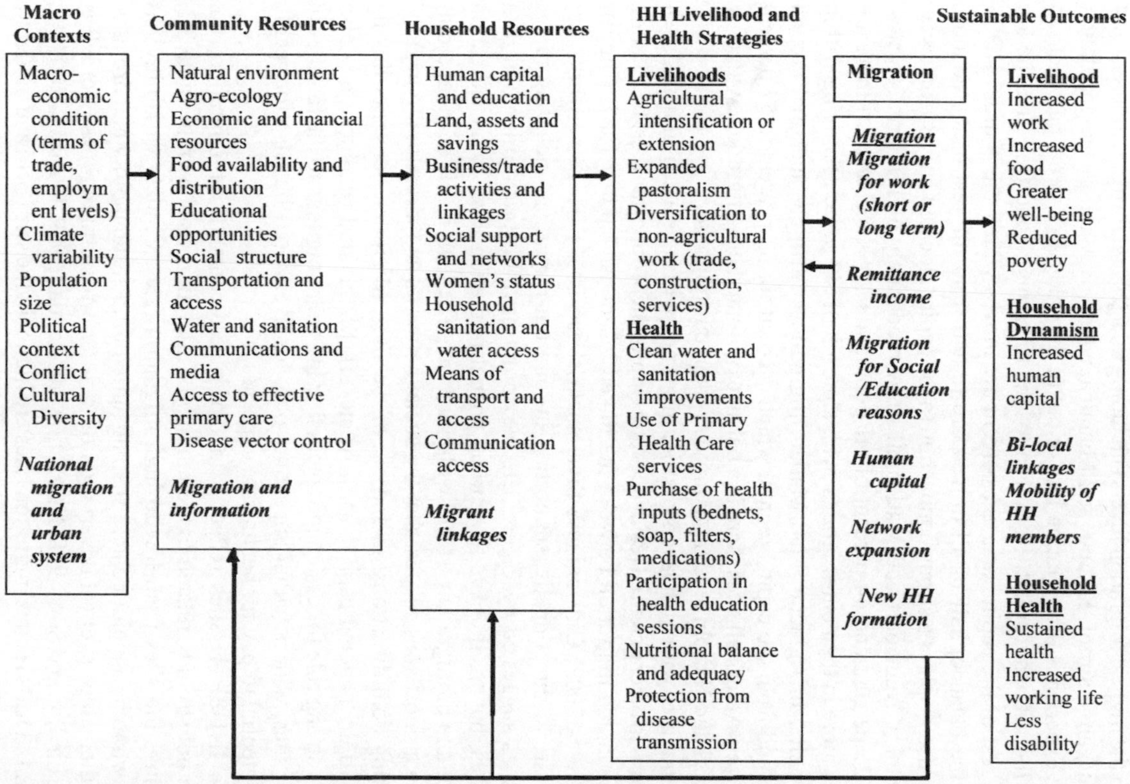

Figure 3.1 **Integrated model of sustainable livelihoods, migration and health**

Table 3.1 shows the sites ranked by level of migration, as measured by the percent migrant in the population. The sites exhibit considerable variability in their prior migration experience. In all but two sites, migration experience is very high, where one out of three or four individuals have some migration experience. The sites with the highest migration levels (including all ages and forms, temporary and permanent) are Nairobi, Kisumu and Manhiça, where over half of the population has some experience of migration into or out of the study area in the past year. In Nairobi, the national capital, 16 percent of the migrants to and from the site are from the rest of the capital area, 31 percent are moves within the Nairobi HDSS site and the balance, half of all moves, are to and from the rest of the country. Manhiça also displays a strong connection to the capital, located only 80 km away. Kanchanaburi, Agincourt and Matlab all show moderately high migration rates, between 27 percent and 44 percent of the population. At Kanchanaburi there are two times more in than out-migrants, with almost half (47 percent) moving within the study site. In Agincourt and Matlab migration is strongly anchored around moves to and from the capital region, which account for 33 percent to 57 percent of all moves. Across all sites there is a balance of in- and out-migration, with only Kanchaburi and Nairobi tending to more in- than out-migration.

Communities with high levels of prior migration will have a denser network of connections to other potential residences and employment opportunities, facilitating and perpetuating migration between the linked locations (Findley 1987, Cliggett 2000, Castles 2002, Gidwani and Sivaramakrishnan 2003). These networks can also operate bi-directionally, supporting in- and out-migration, including return migrants, as suggested by Figure 3.1. We do not have historical migration data for all sites, but we do have the current annual migration level as an indication of the volume of migration 'stock' within the site. The high levels of in- and out-migration at most sites suggest that the social networks and migration 'corridors' are working both directions, in and out of the sites. Migration networks linking the sites to the national capital, as in Agincourt, Matlab and Manhiça, may also indicate a strong role for cash remittances flowing from migrants to the families in the study sites, with potential impact not only for livelihoods but also for health, via the potential to pay for health care expenses.

These INDEPTH sites reflect a mix of macro-contexts for migration, but these are not strongly related to the ranking of migration, as shown in Table 3.2, which ranks the sites from low to high migration rates. Climate variability has been shown to be a major context shaping agricultural sustainability and drought-prone regions face cyclical devastation of their agro-pastoral regimes (Cordell et al. 1996, Findley and Sow 1998, Hampshire and Randall 1999, Konseiga 2007). In the seven INDEPTH sites migration rates are higher in the semi-arid and moderate tropical zones. Both Agincourt and Manhiça have high susceptibility to drought and hence to drought-related migrations. Additionally, when the agricultural cycle is highly seasonal alternating between cultivation season and dry or fallow seasons, the conditions are ripe for circular migration (Hugo 1998). This seasonal influence on migration is evident in these sites, to the extent that the observation

Table 3.1 Migration geography of INDEPTH sites

Sites	Migrant	In-Migrant	Out-Migrant	Local Migrants	Migrations to/ from Capital
Filabavi, Vietnam	13%	6%	7%	21%	6%
Matlab, Bangladesh	27%	13%	14%	27%	57%
Agincourt, South Africa	35%	17%	18%	5%	33%
Kanchanaburi, Thailand	44%	27%	17%	47%	11%
Kisumu, Kenya	58%	30%	28%	NA	NA
Manhiça, Mozambique	60%	32%	28%	8%	38%
Nairobi, Kenya	75%	42%	33%	31%	16%

Note: Migration rates are per 100 persons (total population), single year rates.

of seasonal migrations was possible through the schedule of surveillance relative to the timing of seasonal labour migrations. Two sites with pronounced seasonality in their agricultural cycles alternating between pronounced wet and dry seasons, Kisumu and Manhiça, have high migration rates, but Agincourt, which also has pronounced agricultural seasonality, has only moderate migration rates. In the three Asian sites (Kanchanaburi, Matlab, Filabavi), cultivation is continuous throughout the year with two crops produced and two out of three of these sites have low migration rates.

Local variations among communities shape the resources and potential livelihood strategies available to families, but there is no consistent relation between subsistence agriculture and migration. The six rural INDEPTH sites vary in their mix of subsistence agriculture and non-agricultural income source. As seen in Table 3.2, Filabavi has a majority (81 percent) of dependent on subsistence agriculture, yet it also has the lowest migration rates. This could be reflective of a higher level of agricultural productivity in Filabavi, but also could reflect the fact that the very poorest may be less migratory due to insufficient resources to cover the costs of migration (Hugo 1998). The relation is seen more strongly at Manhiça and Kisumu, where half (51–53 percent) of the families are subsistence farmers and they have some of the highest migration rates across all sites. At Matlab, Agincourt and Kanchanburi 25 percent or less are subsistence farmers and they have low to moderate migration rates.

An essential complement to the agro-pastoral strategies are opportunities to generate wage income, to make up for regular shortfalls between agricultural income and subsistence income or as a hedge against crop or pastoral failure in the short or long-term. Where access to non-farm or wage labour opportunities are limited in a community, families have often resorted to migration to fill the

Table 3.2 Environmental and agricultural context of the INDEPTH sites

Sites	Climate zone	Agricultural Seasons	Climate Variability	Subsistence	Non-Agricultural Income
Filabavi	Tropical	2 crops	None	81%	19%
Matlab	Subtropical	2 crops	Floods	28%	43%
Agincourt	Semi-arid	Wet-dry	Drought	17%	77%
Kanchanaburi	Tropical	2 crops	None	20%	80%
Kisumu	Moderate tropical	Wet-dry	None	58%	42%
Manhiça	Semi-arid	Wet-dry	Drought	51%	20%
Nairobi	Moderate tropical	None	None	1%	99%

gap in their income needs (Findley 1977, Stark and Yitzhaki 1988, Massey 1990, Payne and Lipton 1994, Findley et al. 1995, Cordell et al. 1996, Findley and Sow 1998, Hugo 1998, Hampshire and Randall 1999, Ezra 2000, Konseiga 2007). If this pattern held, then sites with high levels of migration would also have low proportions of households with non-agricultural income, but there is no regular relation between proportion with non-agricultural income sources and migration at these sites. Among the rural sites, Kanchanaburi, Agincourt, Matlab and Kisumu have high levels of participation in non-agricultural activities, with from 42 percent to 80 percent of the households reporting a non-agricultural source of income and they also have moderate to high migration rates. At Manhiça the low level of participation in the cash economy is paired with a high migration rate, while the opposite is found at Filabavi. As will be discussed in the forthcoming chapters, one reason for lack of a relation between migration and access to non-agricultural income sources is that the non-agricultural income sources are located both in and outside of the local community, so that both migrants and non-migrant households may have access to non-agricultural income sources.

The macro-economic context of all sites is one of poverty. They are located in countries ranked below the median of all countries in their Human Development Index. As shown in Table 3.3, Thailand and Vietnam are closest to the median, ranked 81 and 114 out of 179 countries, with scores of 0.786 and 0.718, respectively. All others are lower, in the following order: South Africa (HDI=0.670), Kenya (HDI=0.532), Bangladesh (HDI=0.524) and Mozambique (HDI=0.366). With the exception of Mozambique, these moderate rankings suggest that nationally there is a moderate level of opportunity. The Manhiça site in Mozambique, which has some of the highest migration rates of all the INDEPTH sites, is situated in one of the more impoverished countries of the world, suggesting strong economic forces behind out-migration, including to neighbouring countries with more economic opportunities.

Table 3.3 Macro-economic macro-context of the INDEPTH sites

	Nation, 2006			Poverty		Adult Literacy		School Enrolment	
Sites	Rank	HDI	GDP (PPP$)	National	Site	National	Site	National	Site
Filabavi	114	0.718	2,363	13%	56%	90%	99%	62%	33%
Matlab	147	0.524	1,155	37%	58%	53%	45%	52%	94%
Agincourt	125	0.670	9,087	23%	43%	88%	73%	77%	68%
Kanchanaburi	81	0.786	7,613	9%	17%	94%	NA	78%	5%
Kisumu	144	0.532	1,436	31%	82%	74%	51%	60%	97%
Manhiça	175	0.366	739	48%	NA	44%	47%	55%	NA
Nairobi	144	0.532	1,436	31%	59%	74%	60%	60%	48%

Source: National HDI and national statistics, UNDP, Human Development Index, 2006, Country Fact Sheets 2008/2009. http://hdrstats.undp.org/2008/countries/country_fact_sheets.

Notes: Site data drawn from the information in the chapters complemented by the following sources; INDEPTH Network, 2005 *Measuring Health Equity in Small Areas: Findings from Demographic Surveillance Systems*. Aldershot: Ashgate. % Poverty = Quintile 1–3 in poverty rankings; Otherwise, % Poverty derived from statistics in the chapters or in the Health Equity book. Poverty rate for Nairobi is % of HH without any toilet access. If not listed, literacy rate estimated as all with secondary or higher education plus half of those in primary, assuming that four years needed for literacy.

Migration within the country, which makes up a large share of migrations at these sites, is likely to be more strongly influenced by relative socio-economic disparities with the country. As seen in Table 3.3, the INDEPTH sites all have higher poverty rates than in the national setting. All the sites have poverty rates at least 20 percent higher than the national rate. Although the poverty measures are not exactly comparable, since the poverty measures of the INDEPTH sites are based on a level of living index based on consumer durables, the magnitude of the differences suggest that the INDEPTH sites have significantly higher poverty levels, which would translate in greater need to explore migration or other opportunities to generate additional income.

There are also disparities in human capital. As seen in Table 3.3, at all but two sites for which we have adult literacy data, the adult literacy rate at the INDEPTH sites is well below the national literacy level. The greatest disparities are found in Kisumu and Agincourt, where adult literacy lags behind the national rate by 25 percent or more, making migrants from this region less competitive in the national labour market and more likely to leave to pursue agricultural or non-skilled wage labour jobs. The remaining sites have literacy rates closer to the national average, both above and below, where the residents of the region would be roughly competitive with those of others in terms of human capital, implying that their choices might be broader than in the other three sites.

School enrolment differs which could further shape migration. The higher the school enrolment (rates are all school ages combined), the more likely that there will be people with skills to create and sustain more diversified livelihood options without migrating. At the same time, when migrating they would be more likely to seek more diverse labour markets where there are potential returns to a higher level of human capital. Matlab and Kisumu are sites with school enrolment levels at or above the national level, where we would be likely to see out-migration of persons with more schooling.

When the level of living is low in a community, there could be pressure to move to communities with higher levels of living, either permanently or in order to earn money to send back home to invest in community improvements (Findley and Sow 1998). Conversely, sites with relatively high standards of living, with access to safe water, toilet facilities and electricity, there would be less pressure to move. This is the relation seen at these seven INDEPTH sites. As shown in Table 3.4, the sites where most families have access to these basic infrastructure services, Filabavi, Matlab, Agincourt and Kanchanaburi, all are sites with low to moderate migration rates. In these sites, not only does the electricity open opportunities for the families, electricity service contributes to the sustainability of mechanized additions to agriculture (for example, milling and processing) and to the establishment of industrial or other productive enterprises, supporting diversification of the economy, which is partially born out by the high proportion of households with non-agricultural income in two of these three sites (see Table 3.2). Thus, compared to the other rural sites, these might be ones where there could be less out-migration and more in-migration and circulation. Nairobi's living standards reflect the more urban context where most families report having a radio or television (mostly, a radio) and permanent flooring to their home. Those with the lowest average level of living are Matlab, Kisumu and Manhiça, all of which have very limited penetration of communication media (radio or television) and low housing quality (packed dirt floors, limited access to water or toilets/latrines). These are also the sites with the highest migration rates.

In addition to these contextual economic underpinnings for migration to and from the INDEPTH sites, the timing and pattern of migration is further shaped by location, political and social factors. As seen in Table 3.5, Nairobi stands out as a site where sporadic conflict in the country continues to propel migrants to the site. Three of the sites (Kanchanaburi, Agincourt and Manhiça) are destinations for refugees from conflict in neighbouring regions or countries and the migration flows to these sites were heavily influenced by those conflicts. These sites have a mixture of natives and refugees, which can continue to influence opportunity structures and potential onward migrations among the refugees, either for themselves or their children to other sites in or out of the study area where they can access land or jobs.

Distance to urban labour markets affects migration propensity and the pattern (White and Lindstrom 2005). The INDEPTH sites are located at varying distances from the capital city. Those within 150 km of the national capital (Kanchanaburi, Matlab, Nairobi, Manhiça, Filabavi) are located in areas where it is more likely that there will be regular transport linking them to the capital, facilitating circulation. In addition, these sites, except Nairobi, which is the capital, might have higher out-migration rates to the capital cities because of the greater access to information from the capital via visitors, radio and trade. This would be particularly likely for youth seeking additional education or opportunities to build a livelihood around non-farm employment. The other sites will have migration patterns that are more affected by access to regional or district capitals or involving rural-rural migration, including circulation, to areas offering primary sector employment opportunities.

The timing and pattern of migration is also influenced by social customs regarding marriage and family formation. Where marriage customs require a new wife to move to her husband's village and/or for that to be a village other than her own, many women will move at the time of their marriage (late teens to early twenties). This is the case for Matlab. However, in virtually all the sites migrations are also anticipated when a couple begin to have a family. As will be discussed in the chapter on age-profiles of migration, this affects the migration not only of women but also of their children who accompany their mothers.

Overall migration rates are also affected by the underlying demographic structure of the community. Communities with high child dependency rates are likely to have higher out-migration and circulation rates, as parents seek additional income to support their family, as in the case of Manhiça. Given the social division of labour within a household, unless the mothers leave children with aunts or their mothers and/or can take their children with them, it is likely in these cases that male migration would dominate over female migration. Out-migration rates may

Table 3.4 Community living standards at INDEPTH sites

Sites	Radio or TV	Permanent floor	Safe water	Toilet access	Electricity
Filabavi	50%	88%	71%	93%	93%
Matlab	42%	7%	95%	76%	60%
Agincourt	76%	96%	95%	67%	89%
Kanchanaburi	50%	88%	71%	93%	93%
Kisumu	50%	23%	16%	10%	9%
Manhiça	35%	45%	46%	55%	2%
Nairobi	68%	90%	5%	41%	7%

Table 3.5 Situational and social factors affecting migration timing

Sites	Conflict (Now or Past)	Distance to capital (km)	Marriage Migration	Move to start family	% Pop 15-24 years	Child Dependency
Filabavi	No	60	No	F	21%	0.72
Matlab	No	50	Yes	F	51%	0.86
Agincourt	Yes	500	No	F	24%	0.70
Kanchanaburi	Yes	129	No	M&F	12%	0.69
Kisumu	No	500	No	F	43%	0.85
Manhiça	Yes	80	No	F	19%	1.34
Nairobi	Sporadic	5	No	F	60%	0.81

Note: Sporadic Conflict coded as 'Yes'.

also be higher from communities with a high proportion of youth 15–24, the prime migration ages, as is seen in Nairobi and Kisumu.

Implications of Community Context for Livelihoods and Health

Community context, livelihood options and migration interactively affect each other. In this chapter, the focus has been on how different dimensions of community shape context. As documented in Tables 3.2–3.5, several of the key factors summarized in Figure 3.1 appear to have a role in shaping overall migration patterns at these sites. The sites in semi-arid or moderate tropical zones, subject to climate variability and having sharply demarcated agricultural seasons tend to have higher migration rates. With the exceptions of Filabavi and Nairobi, at opposite ends of the spectrum, the overall site migration rate rises with dependence on subsistence agriculture. Poverty is a major force shaping migration and the strongest relations between poverty and migration are seen in the higher rates of migration in the communities with lower living standards. At least at the community level, other situational and social factors appear to have less influence, although a history of conflict in the region is associated with high migration rates at four sites.

These contextual influences can be seen as setting the stage for the complex interplay of community, household and individual forces which ultimately shape migration into and out of the INDEPTH sites. The analyses in this chapter are static, based on the characteristics of the sites at one point in time and do not capture the dynamic changes that iteratively influence households and their migrations. The process of contextual influence is circular, in that migrations introduce changes to the opportunity space for the families of the community, both currently and in the future. Just as the previous migration patterns set up networks that enable subsequent migrations, the remittances and information brought back by the migrants can shape subsequent developments of physical, economic, human or social capital, such as through introducing irrigation systems, construction of clinics or schools, establishment of trade networks or new businesses, strengthening of micro-credit savings groups or simply infusion of information and ideas from the returned migrants.

Through the unique power of demographic surveillance over a long period of time, the studies reported in this volume are able to portray these dynamics, showing how over time migration impacts community socio-economic structure, reducing household vulnerability. By focusing on how households with migrants change relative to those without migrants, the analyses presented in this volume shed new insights on the cumulative changes associated with migration. In their chapter in this volume, Collinson and co-authors document the progressive impact of migration on raising the levels of living over a five year period in Agincourt. In their chapter on migration and livelihoods in Matlab, Alam and colleagues examine the educational attainment of children, comparing migrants to non-migrants in Matlab. In their analysis, Punpuning and Guest explore the ways that migration impacts on family's farming patterns in Kanchanaburi.

Part of maintaining household livelihoods is maintaining the health of the residents and as with economic survival, health maintenance is influenced by a number of community factors, as described in Figure 3.1. As seen in Table 3.4, at least three of the INDEPTH sites have community contexts which are challenging for sustaining child health, due to limited access to toilets or latrines or clean, safe water. Further, adult literacy levels are below 50 percent at three sites and poverty and economic disparities are great at all sites. In such circumstances, just as migration might improve economic chances for families, migration might also be instrumental to improving health outcomes for the children in the migrant households, through the introduction of greater human capital (skills and education), social capital (social networks and social support for health promotion activities) and economic capital (remittances and money brought back). The chapters from Nairobi, Filabavi and Kisumu all examine the relation between migration and child health. The Nairobi and Filabavi chapters make an interesting contrast, because while both focus on children of migrant parents, Nairobi focuses on risks to child health experienced by the children of recent in-migrants, while the Filabavi chapter focuses on how out-migration of mothers versus fathers differentially impacts the health of the children left behind in the rural area. The Kisumu chapter serves as a kind of bridge between them, focusing on the survival chances of in-migrant children, contrasting children from rural versus urban areas. The consequences of migration, especially those who leave and return, for adult mortality are explored at Manhiça. The assessment of the impact of migration on adult health is complicated by the influence of exposures to diseases while the individual is out-migrated and thus it is not clear how migration impacts the long-term health of those who go and return. As will be seen in their discussion, the community context affecting health care access is critical to understanding the impact of migration on health.

Taken together, the chapters in this volume will provide a rich accounting of community migration, livelihood and health interactions from the uniquely dynamic perspective of the INDEPTH sites. While we have only been able to give you a snapshot view of the community context in each of these sites, in the accounts below the transformative impact of migration on community will be illuminated, as the authors discuss the dynamics of migration at each of their sites.

References

Adekanye, J.B. 1998. Conflicts, loss of state capacities and migration in contemporary Africa, in *Emigration Dynamics in Developing Countries*, edited by R. Appleyard. Aldershot: Ashgate, 165–206.

Amuedo-Dorantes, C. and Mundra, K. 2007. Social networks and their impact on the earnings of Mexican migrants. *Demography*, 44(4), 849–64.

Bilsborrow, R.E., Richard, E. et al. 1991. *Land Use, Migration and Natural Resource Deterioration: The Experience of Guatemala and the Sudan.* New York: Population Council and Oxford University Press, 125–47.

Castles, S. 2002. Migration and community formation under conditions of globalization. *International Migration Review*, 36(4), 1143–59.

Cliggett, L. 2000. Social components of migration: experiences from southern province, Zambia. *Human Organization*, 59(1), 125–35.

Cordell, D.D., Gregory, J.W. and Piche, V. 1996. *Hoe and Wage: A Social History of a Circular Migration System in West Africa.* Boulder: Westview.

Ellis, F. 1999. Rural livelihood diversity in developing countries: evidence and policy implications. *Natural Resource Perspectives*, 40, 1–10.

Entwistle, B. 2007. Putting people into place. *Demography*, 44(4), 687–703.

Ezra, M. 2000. Leaving home of young adults under conditions of ecological stress in the drought-prone communities of northern Ethiopia. *Genus*, 56(3–4), 121–44.

Findley, S.E. 1977. *Planning for Internal Migration: A Review of the Issues and Policies in Developing Countries*. International research document No. 4, International Statistical Programs Centre, U.S. Bureau of the Census. Washington DC: Government Printing Office.

Findley, S.E. 1987. An interactive contextual model of migration in Ilocos Norte, the Philippines. *Demography*, 24(2), May.

Findley, S.E. 1991. Towards a contextual model of the health transition, in *The Health Transition: Methods and Measures*, edited by J. Cleland and A. Hill. Canberra: Health Transition Centre of The Australian University, 381–406.

Findley, S.E. 2001. Compelled to move: the rise of forced migration in Sub-Saharan Africa, in *International Migration into the 21st Century*, edited by M. Siddique. Cheltenham and Northampton, MA: Edward Elgar, 275–310.

Findley, S.E. and Sow, S. 1998. From season to season: agriculture, poverty and migration in the Senegal river valley, Mali, in *Emigration Dynamics in Developing Countries Volume I: Sub-Saharan Africa*, edited by R. Appleyard. Aldershot: Ashgate.

Findley, S.E., Traore, S. et al. 1995. Emigration from the Sahel. *International Migration*, December (3/4).

Gidwani, V. and Sivaramakrishnan, K. 2003. Circular migration and the spaces of cultural assertion. *Annals of the Association of American Geographers*, 93(1), 186–213.

Gurak, D. and Caces, F. 1992. Migration networks and the shaping of migration systems, in *International Migration Systems: A Global Approach*, edited by M. Kritz, L. Lim and H. Zlotnik. Oxford: Clarendon Press.

Hampshire, K. and Randall, S. 1999. Seasonal labour migration strategies in the Sahel: coping with poverty or optimising security. *International Journal of Population and Geography*, 5(5), 367–85.

Hemer, P. 2008. Piot, personhood, place and mobility in Lihir, Papua New Guinea. *Oceania*, 78(1), 109–25.

Hugo, G. 1998. Migration as a survival strategy: the family dimension of migration, in *Population Distribution and Migration*. New York: United Nations Division of Social and Economic Affairs, Population Division, 139–49.

Kazemipur, A. 2006. The market value of friendship: social networks of immigrants. *Canadian Ethnic Studies*, 38(2), 47–71.

Konseiga, A. 2007. Household migration decisions as survival strategy: the case of Burkina Faso. *Journal of African Economies*, 16(2), 198–223.

Lin, N. 1999. Social networks and status attainment. *Annual Review of Sociology*, 25(1), 467–87.

Massey, D.S. 1990. Social structure, household strategies and the cumulative causation of migration. *Population Index*, 56(1), 3–26.

Massey, D.S., Arango, J. et al. 1998. *Worlds in Motion: Understanding International Migration at the End of the Millennium*. New York, NY: Oxford University.

Milazi, D. 1998. Migration within the context of poverty and landlessness in Southern Africa, in *Emigration Dynamics in Developing Countries*, edited by R. Appleyard. Aldershot: Ashgate, 145–64.

Payne, P. and Lipton, M. 1994. *How Third World Rural Households Adapt to Dietary Energy Stress: The Evidence and the Issues*. Washington, D.C.: International Food Policy Research Institute.

Sana, M. and Massey, D.S. 2005. Household composition, family migration and community context: migrant remittances in four countries. *Social Science Quarterly*, 86(2), 509–28.

Scoones, I. 1998. *Sustainable Rural Livelihoods: A Framework for Analysis*. IDS Working Paper. I.F.D. Studies. Brighton: University of Sussex.

Settles, B.H. 2001. Being at home in a global society: a model for families' mobility and immigration decisions. *Journal of Comparative Family Studies*, 32(4), 627–45.

Shah, N.M. and Menon, I. 1999. Chain migration through the social network: experience of labour migrants in Kuwait. *International Migration*, 37(2), 361–82.

Stark, O. and Yitzhaki, S. 1988. Labour migration as a response to relative deprivation. *Journal of Population Economics*, 1(1), 57–70.

White, M. and Lindstrom, D. 2005. Internal migration, in *Handbook of Population*. Dordrecht: Kluwer, 307–42.

Chapter 4
Age–Sex Profiles of Migration: Who is a Migrant?

Mark A. Collinson

Introduction

Certain age groups, like primary school children, show very low levels of migration, while young adults in their prime are the most likely to be migrants (Bell et al. 2002, Long 1992, Rogers and Castro 1981). Like mortality and fertility behaviours, regularities in the age–sex pattern for migration reflect underlying biological, psychological and socio-cultural forces that govern behaviour. The age distributions for these demographic behaviours have been mathematically modelled, but these models are being continually re-assessed against new observations to determine if contemporary populations conform to the previously observed standard patterns. In 2004 INDEPTH made a contribution with respect to standardised mortality schedules, that is, model life tables. By computing the age-profiles of mortality in a range of Health and Demographic Surveillance System (HDSS) sites they produced a series of model life tables that better represent the experience of sub-Saharan African populations, taking into account the effect of the HIV/AIDS epidemic (Ngom and Bawah 2004). At the time of writing the INDEPTH Network was also working on a multi-site fertility study. The chapter presented here is the first attempt to examine multi-site migration age–sex profiles in the INDEPTH network

The regularity of age-patterns of migration has been long observed and model migration schedules have been developed since the 1980s. A landmark study was undertaken by Rogers and Castro, in Austria, published in 1981 (Rogers and Castro 1981). They examined 500 age profiles and presented conclusive evidence of regularity in the age pattern of migration. The modal peak of the migration age distribution was centred on the 20–24 age group and was linked to labour force migration. Migration flows were termed 'young' when the peak centred in the 15–19 age group and conversely, 'old' if it centred on ages 25–29 or 30–34. According to Rogers, 'Age-specific rates of internal migration exhibit remarkably persistent regularities in age profiles. These regularities seem to hold all over the world and across time' (Rogers 1988). The mathematical equation that represents the age-profile is a multi-exponential model, with nine parameters used to fine-tune the model to fit a particular empirical curve. The models were expanded to include more parameters in particular to accommodate the range of retirement curves that occurred in the populations studied. Nevertheless, the regularity of the empirical data supported the mathematical model, with the uniform feature being a uni-modal pattern showing young adults in their prime being most likely to migrate, with children accompanying parents and very little migration from middle age onward except perhaps for a retirement spike. The age patterns presented in the data had excesses or deficits of migration in particular age groups in particular settings, but the pattern was uniform enough to be modelled. Work in less developed parts of the world has confirmed the match between the model migration profiles and empirical data, for example in Ethiopia (Berhanu and White 2000).

Raymer and Rogers subsequently used four model migration schedules to model 196 empirical migration flows. They showed the extent to which the multi-exponential model fit these additional migration observations. The authors showed how the age-pattern was stable over time in the United States, using consecutive censuses to plot the age–sex profiles of foreign-born US interregional migrants between 1950 and 2000 (Rogers and Raymer 1999). They conjectured that model migration schedule may be temporally stable in other settings too.

In 1992, Long compared data on migration from nine countries, defining migration as 'a change in residence' recorded in a census or national survey (Long 1992). The age-profiles encountered were very similar to the profiles described above, with a labour curve centred uniformly on age 20–24. The levels of migration intensity varied with New Zealand having the highest and Ireland the lowest. The extent of children accompanying adults in the migration varied between countries, with the United States showing the highest percent of children moving. This corroborated the regularity of age-profiles, but also brought attention to a critical issue. When comparing migration data from different countries the scale of the migration-defining boundary influences the distance needed to cross the boundary. Moves can be considered within boundaries (whether the boundary is a district, a county, a state or a country) or across boundaries, which affects the relative distribution of migration rates and rates of local mobility. The point is that cross-national comparisons of migration need to use similar spatial scales when defining migration. On this issue, the HDSS study boundaries are very close, though not identical in scale. This compatibility of scale is an advantage of the INDEPTH comparison.

Other key scientific work on the international comparison of internal migration includes Bell, Rees and colleagues (Bell et al. 2002, Rees et al. 2000). In addition to supporting the above argument, they stress the importance of comparable indicators. Migration intensity is a central construct which is computed for particular populations using the optimum indicator that can be constructed from available data. The data structure, for example cross-sectional, retrospective longitudinal or prospective longitudinal, influences which comparable migration intensity indicator can be computed. Censuses use 'transitions', that is, changes in residence in the population as the numerator of the indicator. Longitudinal registration systems use migration events, that is, the actual moves made by migrants. These events can be date-stamped which enable longitudinal methods of analysis. Bell and Rees mostly use transition-based data for measures of migration intensity, but they acknowledge the value of event-based indicators.

The INDEPTH data uses event-based in- and out-migration rates to serve as comparable indicators of migration intensity. The HDSS data are prospectively captured, that is, migration events are recorded as they occur in the study population, from which incidence rates are computed. The comparison in this chapter uses data from only one year, 2002, to provide a consistent temporal focus for the cross-national comparison. The number of events by age, sex and period is computed as a function of the total population with the same age, sex and in the same period. The denominators are the person years experienced by members captured in the population registration system.

In the earlier chapters two perspectives were highlighted that should be borne in mind when making international comparisons of INDEPTH data. Firstly, the definition of migration is not the same in each site. Chapter 2 examines the different definitions of migration in the participating sites and discusses how levels and shape of the curves are influenced by the definition of migration in the HDSS. Secondly, the types of settings represented in the book are highly varied. While the literature asserts that migration age-schedules are stable across space this assertion will be tested in this chapter by inclusion of a range of diverse contexts. As noted in Chapter 3, community context with its powerful social and economic forces shapes the age-profile, from gender norms to the spatial distribution of work opportunities. Nonetheless, regularities in age–sex profile of migrants may transcend the differences in context.

What is an Age–Sex Migration Profile?

Before we look at the comparison of migration rates in the sites this section examines the age–sex profile in one site to introduce the profile itself. It is a useful and ubiquitous analytic device used to examine patterns of migration. The age–sex profile is a compact summary of how migration is actually experienced, a measure of who is most likely to move at what point in their lives and whether and how this differs by gender. The socio-economic forces driving the migration are embedded in the curves, but only age is represented in the profile. The age-profile can be used to guide investigations into age and gender-specific socio-economic forces underlying migration.

The migration rates experienced by the population of Kisumu follow the classical pattern of the model migration schedule. It has a uni-modal distribution with a peak in the early twenties associated with entry into the labour force, higher education or marriage. This age group indicates people migrating in their prime, that is, it is a positively selected sub-group of the population. At older ages the intensity of migration rapidly declines and is near its minimum by age 45. An exception is male in-migration at older ages, which has a sub-mode at age 55 associated with retirement.

The female mode starts at a younger age and, compared to males, women aged 15–19 have a high probability of in- and out-migration. The rate remains high for 20–24-year-olds in both sexes. These young adults are moving for purposes of higher education, working or looking for work and the formation of formal or informal unions. A proportion of the female migration profile is made up of woman moving into the husband's household at the time of union formation, but women also enter the labour market or study away from home.

The male peak is from age 20–24 and is associated with young men leaving for work or education. The in-migration curve for males has a lower peak which is displaced slightly to the right, that is, the average age for male in-migration is higher than the average age of male out-migration. The displacement is associated with men returning home from work-related migration that took them away earlier. After age 30, the likelihood of return in-migration is higher than out-migration. This effect is more muted, but still present in the female curve, that is, the average age for female in-migrants is higher than out-migrants reflecting an out and return movement for work related migration. The displacement is lower for women than men due to the higher prevalence of marital migration for women, which is a more permanent migration.

The age groups under the mode, that is 0–15 years, show a clear pattern of migration intensity associated with each age. The youngest children are most likely to migrate and their level of migration is identical with the level of adults

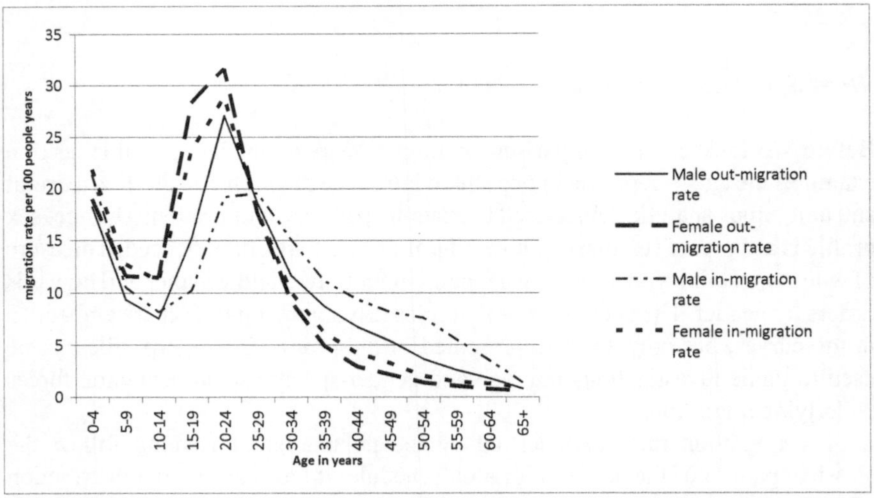

Figure 4.1 Age–sex profiles of in- and out-migration from the Kisumu HDSS site in 2002

Source: Kisumu data, original analysis for this chapter.

aged 25–35 years older. These children are dependent migrants and accompany their parents when they migrate. Therefore, household level factors drive the children's migration and not the age of the child, a status commensurate with their dependency level. Children can move unaccompanied, especially in the African context, for example a foster situation where the care of the child is transferred temporarily to a relative living in a more advantageous situation. The age group 5–14 years is the school-going age and their obligation is to apply themselves to their basic studies. Therefore, levels of migration are at the lowest in this age group. The migration has not disappeared because there is still some older children migrating with their parents, but this tends to be the least mobile age group.

Age–Sex Migration Profiles in INDEPTH Sites

This section presents the comparison of age–sex profiles of migration at the seven participating HDSS sites. All the sites are presented in each graph with four migration flows represented: male out-migration, male in-migration, female out-migration and female in-migration.

Figure 4.2 shows the age-profiles of male out-migration in the seven HDSS sites. The Kisumu curve described above is contained in this multi-country graph, along with out-migration rates from the other sites. The curves from all sites have a similar uni-modal distribution matching the model age-profile for out-migration. The curves are similar with excesses and deficits displayed at different age groups. The peak of the distribution is centred on ages 20–24 for almost all sites regardless of the level. Kanchanaburi has a 'young' distribution with the mode centring on ages 15–19 years. Agincourt has an 'older' distribution with the mode centring on age 25–29. This relates to the site having high levels of circulation among males across the adult working age spectrum, 20–59. The distribution of male out-migrants from Nairobi, Agincourt and Manhiça show a second peak with older men also leaving the site. The age of the second peak varies by setting, with Nairobi and Manhiça peaking at age 49–69 and Agincourt at 40–55. This migration is related to retirement and constitutes men returning home in the case of Nairobi and on-going labour migration in the case of Agincourt and Manhiça. Children may also migrate. The level of child out-migration is variable with Nairobi and Kisumu showing the highest levels. Levels of male out-migration varied widely between sites and in the modal category ranged from 25 percent to 12 percent of the population sub-group.

Figure 4.3 shows the age profiles of male in-migrants into the HDSS sites. While 20–24 remains a key age group there is considerable displacement to the right, with older males moving in compared to those moving out, as shown in Figure 4.2. This suggests that men are returning back to their home places after retirement or for another reason that closes the migration cycle. An outlier to the other distributions is the urban Nairobi site, which has a left displacement compared

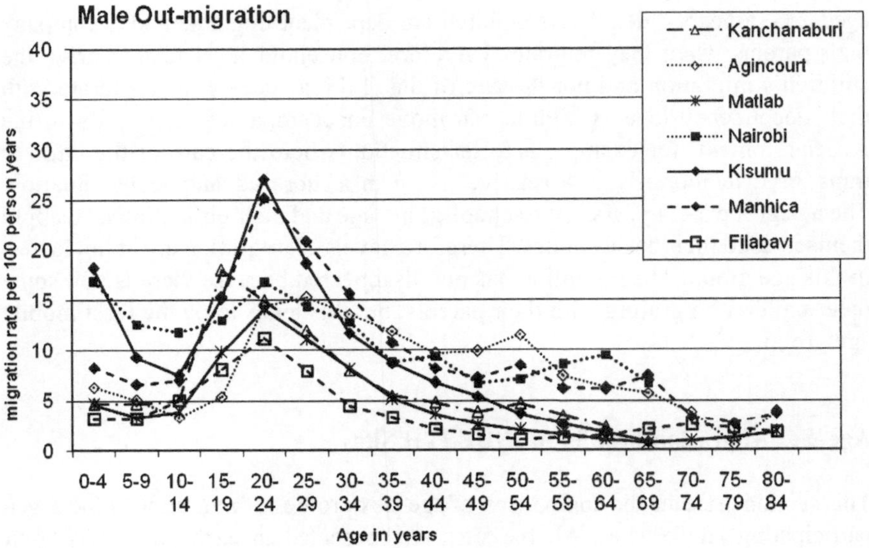

Figure 4.2 Age profiles of male out-migration from seven HDSS sites in 2002

to the out-migrants. Young men are moving into the city for work opportunities, a key migration stream contributing to urbanization. The fact that Nairobi has a 'young' distribution of male out-migrants and the other sites 'old' distributions, reflects that males move for employment or expectation of access to opportunities, which varies between rural and urban settings and is often circular. Urban settings have a more concentrated access to the labour market so the urban sites represent younger men moving in and looking for work, while in the rural sites the older men are moving back from the work place after retirement.

The extent of right displacement of the modal age group varies by site. The most right shifted distribution, centred on ages 30–34, is the rural Bangladeshi site Matlab. Then the rural Agincourt site centred on age 25–29. Three sites show a wide age mode that remains high for men aged 20–29, namely Kanchanaburi, Manhiça and Kisumu. The Agincourt curve is slow to decline and throughout the working ages male in-migration continues. There is a second mode for Agincourt in the 60–64 age group related to retirement. A majority of sites show in-migration due to work retirement in the middle and older adult age groups. For several sites there are still substantial male in-migration rates at ages 50–55 years.

Figure 4.4 shows the age profiles of female out-migration from the participating sites. The pattern continues to prevail that the age group 20–24 is the age when migration is most likely to take place. Sites with a 'young' distribution include Kanchanaburi with female out-migration peaking in ages 15–19. Nairobi has a wider mode that covers the ten year interval 15–24 years. Other sites have high female out-migration rates in the 15–19 age group, but higher yet in the 20–24

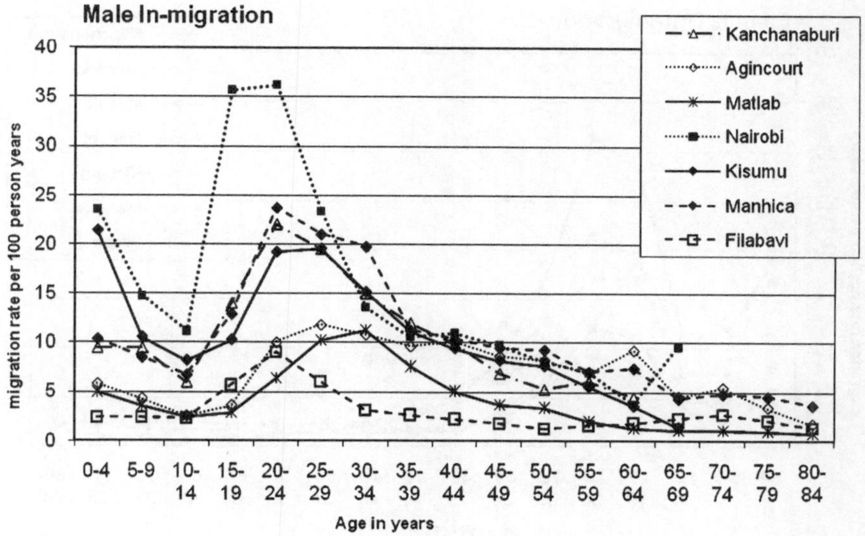

Figure 4.3 Age profiles of male in-migration in seven HDSS sites in 2002

age group. Agincourt has an 'older' distribution with the migration peak centred in ages higher than 24 years. In most sites the rate of migration drops steeply after age 29 years and for many it is at its lowest by age 45, that is, after this age the likelihood of women out-migrating is very low. At the older ages two sites stand out. Manhiça shows a second mode in ages 50–65, related to retirement. In Nairobi, the pattern is different again. Female out-migration remains substantial and does not drop off until age 55 years. There is also a retirement mode in the Nairobi data, but slightly older at ages 60–69 years.

The extent of female children migrating out of the site varies. In some sites like Nairobi and Kisumu the under-five migration rate is equally high with children accompany parents out of the study site. In East Asian sites, like Filabavi and Kanchanaburi, the rate of child out-migration is relatively low. In most sites the age group 10–14 is least likely to migrate.

Figure 4.5 shows the age profile for female in-migration. The modes centre in the 20–24 year age group; however Manhiça has a 'young' profile with the mode for female in-migration occurring in the 15–19 age group. The Agincourt profile again shows an older distribution of in-migration.

There is not much evidence for retirement in-migration, but Nairobi is higher than the other sites with older female migrants leaving the city. There are also elevated levels of older adult female in-migration to Agincourt, Kanchanaburi and Manhiça, all sites with high levels of circulation. The female migration curves do not show much evidence of rightward displacement except in the cases of Agincourt and Manhiça, which show that women are generally less engaged in

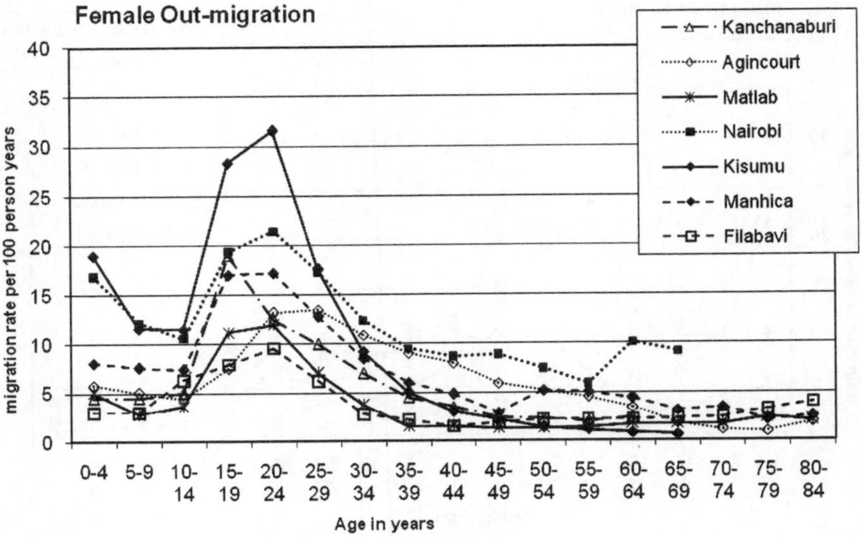

Figure 4.4 Age profiles of female out-migration from seven HDSS sites in 2002

labour migration than man. Migration for women is often of a more permanent nature.

The circular pattern of migration can be examined in the two Kenya sites, because the rural site, Kisumu, is representative of a typical labour-sending community and the urban site, Nairobi, is representative of a typical labour-receiving community. We can examine the circular pattern of migration by looking at the age and sex of in- and out-migration in the two sites. Out-migrants from Kisumu reflect high rates for young adult males and females, as well as for children. In-migrants to Nairobi exhibit high rates for the same groups: young adult male and female, as well as children. Thus, they are mirror images, with the age–sex structure of people leaving Kisumu identical to the age–sex structure of the people moving into Nairobi. Reversing the direction of migration from the city to the rural area helps to highlight the gender differences in circulation. The most likely migrant moving out of Nairobi is a young adult woman and to a less extent a young adult man; along with a high likelihood of child under age five of either sex migrating. As for the reverse pattern, those moving into Kisumu reflect the same age–sex profile Young adult women dominate, with young adult men slightly less prominent in the migration stream. Women in the age group 15–24 possibly with children, leave rural areas for the big city and other destinations. They return in the same age group and also aged 25–34 years, but sometimes older. Men also migrate in these age groups, but more independently of children. Men tend to stay in the city or work destination for longer so the stream of young men out of Nairobi

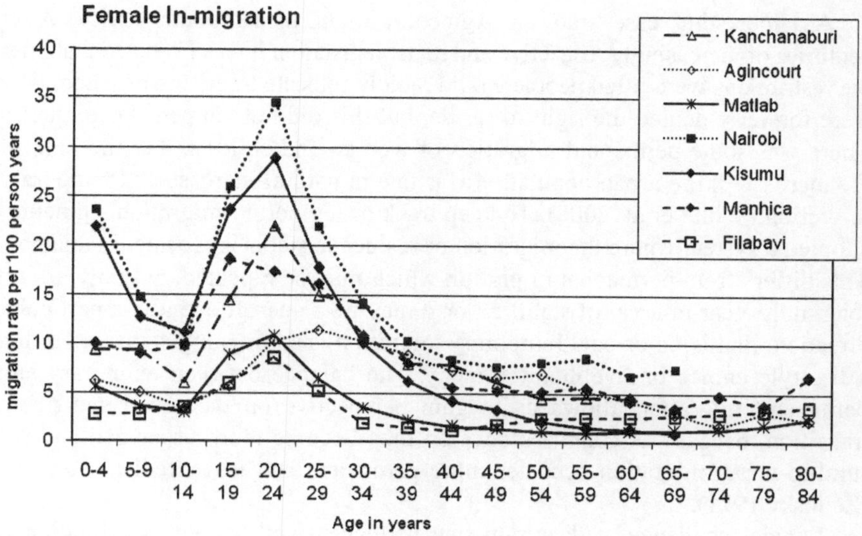

Figure 4.5 Age profiles of female in-migration in seven HDSS sites in 2002

and into Kisumu, does not have the same intensity as it does of young women and children. It is older men who come back from the city and therefore there is a second peak of out-migration of older men from Nairobi.

The Relation Between Migration and Temporary Migration

As noted in Chapter 2, the type of migration observed in any migration study is constrained by the data collection system. Some HDSS data systems keep track of migration by dividing it into permanent and temporary based on the intention of the migrant, that is, whether or not the migrant intends to leave permanently. Permanent migration tends to be for reasons of family formation or break-down or whole households move together to a better situation. Unlike temporary migration, permanent migration may be less likely to provide remittance flows back to the sending household.

The demand for labour and educational opportunities drives people to migrate temporarily and often the migrants are positively selected for various forms of human capital. Temporary out-migration is defined as migration out of the study site while the migrant has the intention of remaining part of the household while away. Temporary migration tends to be for reasons of working, work seeking or accessing better education. Temporary migration has social and economic significance for the households related to remittances and bringing back modern ideas.

An interesting case study is Agincourt in northeast South Africa. After centuries of increasingly restrictive and racist legislation it was expected that after the restrictions were lifted people would rapidly relocate to the towns where they were formerly denied the right to reside, but this did not happen (Posel 2006). There was some permanent migration of women and children, but the bulk of the increase in the urban population was due to natural increase and temporary migration (White et al. 2008). To keep track of temporary migration Agincourt adopted a system where the migration or residence status is repeatedly updated. This differs from permanent migration which may be repeated, but only after a long multi-year interval of stability; or it may be a one-off event in a particular direction, that is, in or out. Temporary migration is an ongoing status reflecting a lifestyle choice or livelihood strategy. The balance between temporary and permanent migration allows us to gain perspective on the urban settlement transition, because as countries develop they become more urbanized and the rural-to-urban migration becomes more permanent and less circular over time (Zelinsky 1971).

A major challenge of discriminating permanent and temporary migration is computing migration rates as presented in this chapter. The analysts need to recombine permanent and temporary migrants into migration categories that are not qualified by the permanence of the move. It turns out that merging the permanent and temporary curves is not as easy as one might think. As a repeatable cycle, the temporary migration produces a rate that is akin to a prevalence rate telling us how much of something exists in the population. A permanent migration on the other hand is a long-duration migration or a one-off event that is akin to an incidence rate telling us how many new cases occur in a population in a given time. In the merge, temporary migration has to be dissected to reflect the exact profile of people moving in or out as if they had been recorded with a migration-defining threshold period of three months. By this definition the following temporary and permanent migrants are included as migrants: all permanent migrants that crossed the boundary of the study site in a given period; people that were not temporary migrants in the year before, but became temporary migrants in the observation year (that is, they are out-migrants); people that were temporary migrants in the previous year but stopped circulating and became permanent residents (that is, in-migrants); people who were circulators but remained in the original household for three or more months in the year, hence they passed the threshold period of residence so could be considered migrants. Migrants picked up by the temporary migration system that are not considered migrants according to the three-month threshold definition are people who live mostly at their place of work but come back to the rural household for three months or less in a year. These people don't accumulate enough time in the rural household to pass the threshold period so they are not considered migrants. Another type of move that does not qualify as migration is if a person migrates out but returns within the three months threshold period. In sum, the categories of temporary and permanent migration can be recombined into migration with a definition based on a threshold period (for example three months) that a person resides in or out of the household.

How Comparable Are the Migration Levels?

The age–sex profiles of migration reveal how the intensity of migration bears upon certain age groups more than others, notably young adults, but there is a diversity of levels presented across the different sites. A question addressed in this section is whether we can determine if some places have higher migration levels than others. There are some key obstacles to making cross-site comparisons and we discuss one in more detail. Firstly, there are structural site-level effects, for example sites that are not a contiguous geographical area can have higher migration levels because it is easier to cross the migration-defining boundary. Secondly, the migration definitions are not comparable between the participating HDSS sites. Threshold periods used to determine migration range from one to six months in the seven sites. Thirdly, there are macro-structural factors that vary enormously between sites, as discussed in the previous chapter. The extent of subsistence agriculture and the proximity to town, level of infra-structure development and migration context all help to determine the level and direction of migration flows.

Despite the contextual and site-definitional differences there is a remarkable similarity in the age structure of the migration. However, we cannot be sure that the levels are comparable due to the definitional incompatibilities, a key one of which is the time threshold or probation period used to define migration. Threshold periods vary by site. The shortest threshold is Kanchanaburi at one month and longest is Matlab at six months. Table 4.1 compares the age–sex structure of in- and out-migration in these two HDSS sites. The site with the higher threshold period has fewer moves classified as migration and thus lower migration rates. The site with the lower threshold period has consistently and significantly higher migration rates. In this comparison Kanchanaburi has higher rates than Matlab, but we cannot claim that the comparison is valid because the migration definition implies a higher rate for the lower threshold site. Thus, despite statistical significance we cannot conclude that Kanchanaburi has a higher level of migration than Matlab.

A contrary finding is displayed in Table 4.2, which shows higher rates of migration experienced in one group of sites compared to another. The table compares the combined migration rates of a group of three sites (Agincourt, Filabavi and Manhiça) with a second group (Kisumu and Nairobi). The first group has a threshold period of three months and the second group a threshold period of four months. Based on the statistical tests comparing migration rates between the groups, the four month threshold sites (Nairobi and Kisumu) have much higher migration rates. The four-month sites should have lower migration levels than the three-month sites, but in reality the opposite holds. In fact, the four-month sites have the highest migration rates for ages less than 65. Therefore, we can say that despite the differences in threshold period, Nairobi and Kisumu had higher levels

Table 4.1 Migration rates by age and gender, by migration duration minimum threshold (one versus six months) (with p values for test of significant differences)

	Age group	Kanchanaburi, with a 1-month threshold period	Matlab, with a 6-month threshold period	P-value
Male out-migrants	0–14	4.8	3.9	0.000
	15–34	13.3	10.8	0.000
	35–64	4.1	2.7	0.002
Female out-migrants	0–14	4.7	3.8	0.000
	15–34	12.1	8.5	0.000
	35–64	2.9	1.5	0.000
Male in-migrants	0–14	8.3	3.7	0.000
	15–34	17.6	7.7	0.000
	35–64	7.5	3.8	0.000
Female in-migrants	0–14	8.3	4.1	0.000
	15–34	16.4	7.5	0.000
	35–64	5.9	1.4	0.000

of migration than Manhiça, Filabavi and Agincourt. The evidence is firmer here than in the above case because the difference in level was not predicted by the migration definitions. This implies that the structural and historical forces are more important than the definitional inconsistency in this comparison. In both Kenya sites the level is higher than the comparison group despite the longer threshold of time taken to define the migration. More cross-country work is needed to determine whether the migration rates really are higher in Kenya than in other parts of the developing world and if so why

Conclusion

The analyses presented in this chapter confirm the regularity of migration by age and sex in the INDEPTH sites, showing conformance to the basic patterns outlined by Rogers and Castro. Even though the INDEPTH sites represent communities and countries with high levels of poverty and socio-economic vulnerabilities, the age–sex profiles of migration are very similar to those found in the more developed countries two to three decades previously. There are some differences, with a broader peak in the age group 15–34. Thus, the seven sites provide further

Table 4.2 Migration rates by age and gender, by migration duration threshold (three versus four months) (p value for test of significant differences)

	Age group	Agincourt, Filabavi, Manhiça with 3-month threshold period	Nairobi and Kisumu with 4-month threshold period	p-value
Male out-migrants	0–14	5.3	12.7	0.000
	15–34	13.1	16.2	0.000
	35–64	6.4	6.6	0.023
	65+	3.3	0.9	0.001
Female out-migrants	0–14	5.6	13.6	0.000
	15–34	10.6	19.7	0.000
	35–64	4.3	5.3	0.003
	65+	2.4	1.2	0.009
Male in-migrants	0–14	5.0	15.0	0.000
	15–34	11.5	21.6	0.000
	35–64	6.6	8.0	0.033
	65+	3.4	1.4	0.001
Female in-migrants	0–14	5.8	15.7	0.000
	15–34	10.2	22.1	0.000
	35–64	4.7	5.7	0.007
	65+	3.4	1.1	0.000

evidence of the universality of the age–sex profile of migrants. The modal group is young adults, sometimes accompanied by children. Labour migration is a key component of these migration profiles, also children accompanying migrant parents and to a lesser extent marriage or marriage dissolution or households moving to a better situation. There is also evidence of circulation for labour, especially among male migrants. Comparing levels is more complicated and better data collection and statistical methods are needed, including standardising the definitions of migration-defining boundaries and threshold periods.

References

Bell, M., Blake, M., Boyle, P., Duke-Williams, O., Rees, P., Stillwell, J. and Hugo, G. 2002. Cross-national comparison of internal migration: issues and measures. *Journal of the Royal Statistical Society*, Series A 165(3), 435–67.

Berhanu, B. and White, M.J. 2000. War, famine and female migration in Ethiopia 1960–1989. *Economic Development and Cultural Change*, 49(1), 91–113.

Long, L. 1992. Changing residence: comparative perspectives on its relationship to age, sex and marital status. *Population Studies*, 46(1), 141–58.

Ngom, P. and Bawah, A.A. 2004. *Volume 2: INDEPTH Model Life Tables for Sub-Saharan Africa*. Aldershot: Ashgate.

Posel, D. 2006. Moving on: patterns of labour migration in post-apartheid South Africa, in *Africa on the Move: African Migration and Urbanisation in Comparative Perspective*, edited by M. Tienda, S.E. Findley, S.M. Tollman and E. Preston-Whyte. Johannesburg: Wits University Press.

Rees, N., Bell, M., Duke-Williams, O. and Blake, M. 2000. Problems and solutions in the measurement of migration intensities: Australia and Britain compared. *Population Studies*, 54(2), 207–22.

Rogers, A. 1988. Age patterns of elderly migration: an international perspective. *Demography*, 25(1988), 355.

Rogers, A. and Castro, L.J. 1981. *Model Migration Schedules Research Report -81–30*. Laxenburg, Austria: International Institute for Applied Systems Analysis.

Rogers, A. and Raymer, J. 1999. Estimating the regional migration pattern of the foreign born population in the United States: 1950–1990. *Mathematical Population Studies*, 7(3), 181–216.

White, M.J., Mberu, B.U. and Collinson, M.A. 2008. African urbanization: recent trends and implications, in *The New Global Frontier: Urbanization, Poverty and Environment in the 21st Century*, edited by G. Martine, G. McGranahan, M. Montgomery and R. Fernández-Castilla. London: Earthscan, 301–16.

Zelinsky, W. 1971. The hypothesis of the mobility transition. *Geographical Review*, 61(3), 219–49.

PART II
Migration and Livelihoods

Chapter 5
Migration and Agricultural Production in Kanchanaburi, Thailand

Sureeporn Punpuing and Philip Guest

Introduction

The literature on the economic consequences of migration, especially for those left behind, exhibits considerable debate and confusion (Bilsborrow 1998). Some of the confusion stems from inappropriate comparisons used in assessing consequences. The data available for analysis is typically cross-sectional and does not allow an examination of the delayed impact of migration on households.

Research on the impacts of migration on the productivity and living standards of those households left behind, especially households engaged in agriculture, is one area that exemplifies the confusion noted above. Some argue that migration of productive households members may result in increased labour force involvement of non-migrating households members, especially women (Oberai et al. 1989, Nelson 1992). Others argue that in rural areas there exists surplus agricultural labour and the out-migration of that labour is likely to have little impact on the allocation of household labour. Finally, other researchers claim that remittances from out-migrants can reduce the amount of labour required from household members and can result in some household members exiting the labour force (Makinwa-Adebusoye 1993).

In addition to the linkages that have been made between out-migration and changes in labour inputs of remaining household members, out-migration from agricultural households has also been linked to changes in the types of agriculture that are engaged in by households. A study in northern Thailand, for example, found that out-migration led to a reduction in livestock raising, because migrant remittances provided a substitute for this source of income (Singhanetra-Renard 1992). Other studies have noted a shift from more intensive to less-intensive forms of agriculture after the out-migration of household members (Makinwa-Adebusoye 1993).

Analysis of the effects of migration on the economic activities of household members remaining behind requires information on the resources available to the household. Where ample resources are available, either as part of the endowment of the household or through gains made from migration, the effects of migration can be expected to be different from a situation where migration reduces the resources available to a household. As the effects of migration on household resources may differ over time, it is important to examine the impacts of migration within a longitudinal framework.

Examining the effects of migration on households at the place of origin is difficult because most studies do not include non-migrant households in the study design and hence have no way of evaluating if an observed effect is a result of migration or of some other factor (Guest 1998a).

In this chapter we use longitudinal data at the household level set to explore the extent to which migration from agricultural households is associated with changes in the amount of land used in cultivation and the amount of household labour employed in agriculture.

The Economic Consequences of Migration

From the available research it becomes clear that the vast majority of migrants benefit economically from their moves. Globally and in Thailand, most studies of internal migration show that migrants have higher levels of labour force participation than non-migrants, usually have a job arranged before they move or, if not, spend little time looking for a job and earn much more than they would be able to earn undertaking equivalent work in their areas of origin (Chamratrithirong et al. 1995, Bilsborrow 1998, Guest 1998a, 1998b).

The results of studies on the impact of migration on households of origin are more diverse. Skeldon (1997) argues that rural-urban migration is particularly beneficial as a means of alleviating poverty in rural areas. He notes that remittances from temporary migrants provide rural families with cash incomes that can be used to sustain their rural way of life. Oberai et al. (1989), argue that remittances obtained through migration raise the incomes of poorer households in three areas of India. They provide a very positive view of the effects of remittance. Other researchers argue that while remittances might benefit households with migrants, at the aggregate level the effect is a worsening of the distribution of income in their rural areas of origin, with migration being more likely to occur from the better-off households. These households have the ability to finance migration and disproportionately receive the benefits of increased income derived from remittances.

Stark (1991), however, argues that the effects of remittances on income distribution depends on the stage of development of migration streams. As migration becomes more common the costs of migration are decreased and most households can finance migration of one or more members. The relative contribution to household income of migrants from poorer households is likely to be greater than that from richer households, thus possibly acting to reduce income inequality. There is some support for this view, for example, Makannah (1988) in a study of the effects of remittances on rural development in Sierra Leone, concluded that remittances improved income distribution because of the higher proportion of income derived from remittances by the poorer households.

An argument that occurs frequently in the literature is that as so little of remittance money is spent on direct productive investment and much of it is spent on consumer durables or housing, there is little indirect (and no direct) effect of remittances on economic development at either the household or community level. However, the time frame used in the analysis is important. Glytos (1993) argues that the multiplier effects of remittances are often not identified because they take time to develop. In his analysis for Greece, he notes that remittances have strong effects on employment generation, mainly through the effect of increased spending on consumption. Although this also has some impact on increasing the levels of imports, he demonstrates that this effect was relatively mild. Although first round consumption of remittances was mainly in the 'non-productive' areas of housing and other basic needs, this spending generated employment and domestic investment.

The argument that remittances spent on daily living expenses are unproductive ignores the improvements that such uses can have on building household human capital. Returns to investment in areas of human capital, such as health and education also take time before they contribute to economic well-being. Francis and Hoddinott (1993) have illustrated how, in two locations of Kenya, investment of remittances by migrants in the education of their children took a generation to be realized. Analyses of this sort require the use of historical techniques because of the lengthy period over which the processes operate.

Skeldon (2002) notes that in the literature the relationship between poverty and migration is typically either poverty as a cause of migration or migration as a cause of poverty. He argues that there is little evidence of either of these relationships. Instead he argues for a third perspective: poverty alleviated by migration. He states that the weight of evidence from studies across Asia suggest that: 'mobility enhances economic growth and improves the lot of most, but not all, of the population' (Skeldon 2002: 9). He notes that the poverty reduction effect operates at both the individual level and at the household and societal levels.

Much of Skeldon's argument focuses primarily on the role of income transfers; primarily from migrants in urban areas to their rural origin households. However, there has been far less attention paid to how migration of household members affects economic roles of household members left behind. Several studies have suggested that out-migration can change how agricultural production is undertaken. For example, Joekes et al. (1994) provide an example of how the out-migration of males in an area of shifting cultivation studied in Malaysia reduced the length of fallow. This was a result of the unavailability of men to clear new areas for crops. Oberai et al. (1989), note higher labour force participation of women resulting from out-migration of family members, but argue that this is a positive development in increasing female status. A similar finding, but with a different interpretation, is provided by Nelson (1992) who in her analysis of migration in Kenya concludes that while out-migration by men provided women with more independence, this has come at the expense, of more work and insecurity. Meanwhile, women moving to urban areas were generally confined to marginal occupations in the informal sector. Other studies have also indicated that there might be a loss of the occupational skills of family members who rely on remittances rather than working. For example, Makinwa-Adebusoye (1993) found that agricultural output suffered as families at the place of origin increasingly relied on remittances and ignored agriculture.

Over the last three decades Thailand has been undergoing rapid structural change, with levels of migration increasing and rapid reductions in the percentage of the labour force employed in agriculture. Remittances from migrants have become an important source of household income for rural households. Guest (1998a) uses data from two linked surveys of migration in Thailand to show that remittances provide an important supplement to household income in Northeast Thailand. The uses of remittances have important multiplier effects on the economy, with many of the major items of expenditure, for example construction materials and labour being obtained locally. Guest (1998a) also found that remittances had helped reduce the levels of intra-rural household income inequality. His findings suggest that in the rural Northeast of Thailand remittances contribute significantly towards improving household income. In both surveys the household members responding to the questionnaire were asked to report all sources of income, including that earned from selling crops or household resources, during the previous 12 months. They were then asked to estimate the total household net income from these combined sources. Most respondents, especially those living in rural areas, were very definite about the amounts household members had earned. Remittances made up almost a quarter of all household income, ranging from eight percent of the monthly household income of out-migrant households to over 40 percent for those households that contained both return and out-migrants. There was an average of three migrants in this form of household, compared to only 1.7 in households with only out-migrants and 1.4 migrants from households only with return migrants.

Thai data also suggest that remittances may also change patterns of household production. Guest (1998a) in his study of Thai migration data found that payment of wages was an important use of remittances. This suggests that for some households, hired labour may be used to substitute for household labour that has migrated and hence hiring labour can be considered a productive form of investment of remittances. However, Singhanetra-Renard (1992) notes that in a village in North-eastern Thailand that she studied, traditional handicrafts and certain types of agriculture (the raising of livestock) were no longer engaged in, because of the earnings that were obtained from migration. The results of these studies have to be contrasted however, with reports that the temporary migration of men increases the labour force participation and household management skills of women, as well as makes women more likely to participate in agriculture. The study by Singhanetra-Renard (1992), also found that while there was a decline in traditional occupations, new occupations based on entrepreneurship and capital were generated.

Migration changes the structure of households as well as the resource base. As we can see from the brief literature review above, the effects of these changes on productivity of households of origin are unclear.

Data and Methods

Data from the Kanchanaburi Demographic Surveillance System (DSS) is used for this study. The Kanchanaburi HDSS is operated by the Institute for Population and Social Research (IPSR), Mahidol University. The HDSS annually collects data using a population census for every household and for each individual aged 15 years and over, in approximately 100 villages and census blocks randomly selected from the province of Kanchanaburi in the year 2000.

Kanchanaburi, the third largest of the 76 provinces of Thailand, is located in the western part of the country. The province shares a long border with Myanmar and contains a variety of ethnic groups and migrants, both documented and undocumented. The province is predominately agricultural. Rice and plantation crops are grown, with animal husbandry and fruit cultivation also constituting important economic activities. Levels of out-migration, especially of young adults are high (IPSR 2004).

The province of Kanchanaburi is very diverse in its economic structure and levels of economic and social development. Nationally the province ranks highly on many indicators of development. The province, compared to the other 75 provinces of Thailand, has lower levels of unemployment and higher levels of home ownership and ownership of consumer durables. Average educational levels are at the national average as are household levels of income. However, the percentage of households living in poverty (16.9 percent in 2004) is much higher than the national average (UNDP 2007). The aggregate indicators reflect the highly diverse structure of Kanchanaburi. While the majority of households are engaged in agriculture, this sector encompasses both low income (low land rice and upland crops) as well as higher income areas (plantation crops). At the same time, the province is rapidly industrializing, with industrial development mainly occurring in a strip from the capital of the province through to Nakorn Pathom, a province adjacent to Bangkok. The variation in economic structure, in addition to close travel distance to the economic hub of the nation located in Bangkok and surrounding provinces, is a major factor in high levels of migration that are found in the province.

A central component of the Kanchanaburi HDSS is the annual enumeration of all households in the field site communities. The first enumeration was undertaken in 2000 and the fifth was completed in 2004. The enumeration of households is conducted during the middle of each year, starting from the first of July. The data collected includes population, economic, social and health related information. For data collection and comparative purposes each household from which data are collected is given a unique code. This enables household records to be linked over time. The first four rounds of data are included in the analysis presented in this chapter.

The chapter focuses on agricultural households, which are defined as those that used land for agricultural purposes in 2000. The land used could be owned, rented or available free of charge.[1] Outputs from agricultural production could be sold or consumed. The total number of households available for analysis is 4,955.

The data used in the analysis are drawn from the household questionnaire and the household is used as the unit of analysis. All four rounds of the annual census are combined for the analysis. The descriptive analysis, categorizes households by combining information on whether a usual adult member of the household migrated from the household in the three years of observation (2000–2003) and the timing of the first event of migration from the household. Migration is defined as movement out of the village of residence for a period of at least one month. This definition results in four categories: households in which no out-migration occurred in the three years, households where out-migration first occurred three years before the final census round in 2003, households where the first migration occurred two years before the final census round and households where the first migration occurred the year before the final census round. In the multivariate analysis information on the number of out-migrants is also included.

The main outcomes of interest (dependent variables) in the analysis are:

1. The amount of land that households use for agricultural production
2. The amount of agricultural land used for high labour intensive production (defined in terms of use of land for growing rice and other annual or biannual crops)
3. The amount of land used for low labour intensive agriculture (defined in terms of animal husbandry and crops taking more than two years to mature)
4. The proportion of household members whose main occupation is in agriculture.

In the multivariate models, explanatory (independent) variables include the out-migration status of the household, number of out-migrants; number of in-migrants, demographic composition of the household, levels of household debt and household wealth. The multivariate analysis is used to model the extent to which out-migration status of the household in the period 2000–2003 is related to levels of the outcome variables in 2003, after controlling for initial levels of the outcome variables and changes in values of other independent variables.

1 Unfortunately, data on amount of land used for agriculture is not available disaggregated by the status of land ownership. If such data was available it would be possible to determine if the effect of migration was to reduce levels of use of rented land rather than land owned.

Table 5.1 Percentage distribution of households by out-migration status and year (n=4,555)

Migration Status	Year		
	2001	2002	2003
No out-migration	76.0	57.3	43.5
First out-migration in previous 1 year	24.0	18.7	13.9
First out-migration in previous 2 years	–	24.0	18.7
First out-migration in previous 3 years	–	–	24.0

Note: – signifies migration data not available.

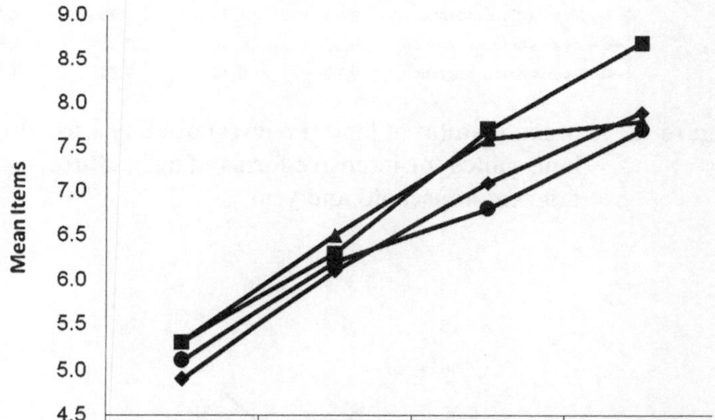

	2000	2001	2002	2003
No Migrant	4.9	6.1	7.1	7.9
Migrant 1 year before	5.3	6.3	7.7	8.7
Migrant 2 years before	5.3	6.5	7.6	7.8
Migrant 3 years before	5.1	6.2	6.8	7.7

Figure 5.1 Mean number of consumer items owned by agricultural households by type of migrant household and year

Results and Discussion

In Table 5.1 the distribution of households by out-migration status is shown. As noted in the previous section, migration status is defined based on the first event of out-migration in the four years of observation available. Between the data collection rounds of 2000 and 2001, 24 percent of households had at least one usual member of the household aged 15 and above move out of the household. The percentage declines for each subsequent period, being 13.9 percent of households

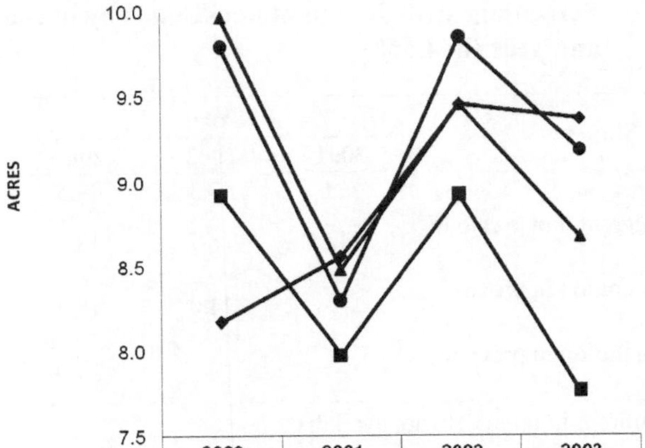

	2000	2001	2002	2003
Non-migration	8.18	8.56	9.46	9.37
1 year before migration	8.93	7.98	8.93	7.76
2 years before migration	9.97	8.49	9.46	8.68
3 years before migration	9.79	8.30	9.85	9.19

Figure 5.2 Mean amount of land (in acres) used by agricultural households for intensive forms of agriculture, by type of migrant household and year

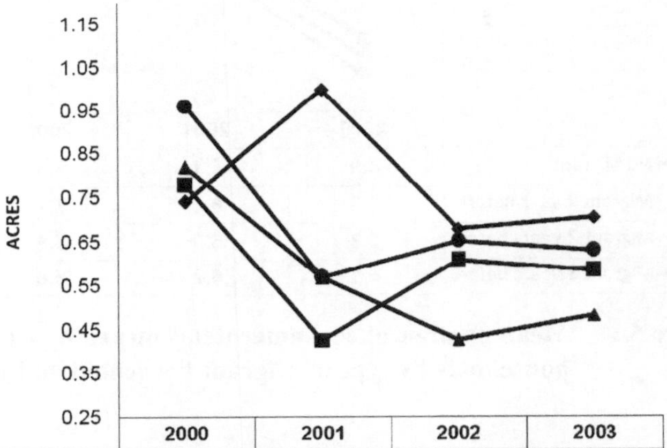

	2000	2001	2002	2003
Non-migration	0.74	1.00	0.68	0.70
1 year before migration	0.78	0.43	0.61	0.58
2 years before migration	0.82	0.57	0.43	0.48
3 years before migration	0.96	0.57	0.65	0.63

Figure 5.3 Mean amount of land (in acres) used by agricultural households for less intensive forms of agriculture, by type of migrant household and year

Table 5.2 Mean number of household members employed in agriculture by year and household migration status (n=4,955)

Migration Status	Year			
	2000	**2001**	**2002**	**2003**
No out-migration	1.73	1.72	1.69	1.70
First out-migration in previous 1 year	1.87	1.93	*2.33*	2.24
First out-migration in previous 2 years	2.06	*2.33*	2.05	2.01
First out-migration in previous 3 years	*2.31*	2.24	2.21	2.15

who, for the three years of observation had their first adult household member migrate between the years 2002 and 2003. It is important to note that the decline across years does not indicate a decline in the amount of out-migration, since those households that had an adult migrant in an earlier year (for example 2001–2002) may also have had a migrant in a subsequent year (for example 2002–2003), however the household is categorized with respect to the first out-migration in the observation period (2001–2002 in the example). It is important to note is that in the three-year period, 56 percent of all households have had at least one out-migrant.

The living standards of agricultural households improved considerably over the four years of observation (see Figure 5.1). These changes occurred irrespective of whether the household did or did not contain migrants. Households without migrants had the lowest wealth, as measured by the mean number of selected consumer durables owned by the household, in 2000. By 2003, their wealth was similar to those households with migrants.

The data shown in Figures 5.2 and 5.3 shows the mean amount of land utilized by agricultural households in intensive agriculture and less intensive agriculture respectively. Although there is considerable annual fluctuation in the amount of land used for intensive agriculture especially in 2001, in part due to availability of water for extra crops, with few villages in our study accessing water from irrigation projects, the results suggest that out-migration is associated with a reduction in the amount of land farmed with intensive crops. In the period 2002–2003, all migration households experienced a reduction in the mean amount of land used for intensive cropping. No decline was observed for households that had no out-migrant in the previous three years.

In contrast, there appears to be very little relationship between out-migration status of the household and the amount of land used for non-intensive agriculture. This suggests that out-migration does not lead to a shift from more to less intensive forms of agriculture. Rather, there appears to be partial withdrawals from agriculture from agricultural households that have members migrate out of the household, with the withdrawal largely a shift from more intensive forms of agriculture.

Table 5.3 Unstandardized ordinary least squares regression coefficients of models predicting amount of land used in agriculture

Variable	Total Agricultural Land		Agricultural Land – Intensive		Agricultural Land – Non-Intensive	
	1	2	1	2	1	2
No out-migration (ref)						
Out-migrant 1 year before	-2.00	-2.71**	-1.84	-2.83**	-0.16	0.00
Out-migrant 2 years before	-0.72	-2.28*	-0.51	-2.29*	-0.21	-0.08
Out-migrant 3 years before	-0.11	-1.84	0.60	-1.69	-0.17	-0.03
Sex ratio 2000	-	-0.00	-	-0.00	-	0.00
Sex ratio 2003	-	0.00	-	0.00	-	-0.00
Dependency ratio 2000	-	0.00	-	0.00	-	-0.00
Dependency ratio 2003	-	-0.00	-	-0.00	-	-0.00
Number HH members 2000	-	-0.17	-	-0.27	-	-0.06
Number HH members 2003	-	-0.30	-	0.25	-	-0.05
Number in-migrants 2000	-	-1.03	-	-1.42	-	0.46
Number in-migrants 2003	-	0.32	-	0.49	-	-0.05
Household wealth 2000	-	-0.11	-	-0.05	-	0.01
Household wealth 2003	-	0.50**	-	0.47**	-	0.05
Household debt 2000	-	0.00**	-	-0.00	-	0.00**
Household debt 2003	-	0.00**	-	0.00**	-	0.00**
Male out-migrant 15-59 2003	-	1.14	-	1.51*	-	-0.16
Female out-migrant 15-59 2003	-	-0.49	-	-0.23	-	-0.12
Land for agriculture 2000	-	0.88**	-		-	
Land intensive agriculture 2000	-		-	0.92**	-	
Land non-intensive agriculture 2000	-		-		-	0.25**
Constant	10.53	-2.87	9.81	-3.02	0.72	0.19
N	4397	4397	4397	4397	4397	4397
R-square	0.001	0.504**	0.001	0.523**	0.000	0.122**

Note: * - significant at .05 level; ** - significant at .01 level. Household wealth is measured through ownership of the following household durable goods: colour television, video/compact disk/digital video disk, stereo, line or mobile telephone, computer, electronic pump, air condition, sewing machine, washing machine, microwave, refrigerator, bicycle, motorcycle, tri-cycle, local agriculture vehicles, car, pickup, truck and small agriculture tractor. Cronbach's Alpha measure of reliability of the index is 0.72. When correlated with a composite index on Principal Component Analysis (PCA), the correlation between the two indexes is 0.97. However, the number of assets is more appropriate because we want to analyse change over time of the same households, while the PCA measures relative wealth i.e. wealth is compared with other households included in the analysis.

The data displayed in Table 5.2 suggest that out-migration results in a significant reduction in the amount of household labour employed in agriculture in the year immediately after the migration. This reduction continues, although in a more muted form, in subsequent years. The mean number of persons employed in agriculture is highest at the start of the year in which migration occurs. In the year out-migration took place, each migrant household employed approximately 2.3 persons in agriculture. Non-migrant households have the lowest mean number of household members employed in agriculture, with the difference in means in the year during which migration took place between the non-migrant households and the households in which a migration took place in that year, being approximately 0.6 of a household member. It is not possible from the data presented in Table 5.2 to determine if migration from agricultural households helps relieve agricultural labour pressure on these households or if the reduction in the number of household members employed in agriculture is a result of the substitution of migrant labour for household agricultural labour.

Land Used in Agriculture

Table 5.3 presents the results of the ordinary least squares regression of the three dependent variables, measured in 2003, regressed against a set of variables measured in 2000 and in 2003. For each of the dependent variables two models are estimated. The first model includes only the variable that indexes the out-migration status of the household. The second model adds all other predictor variables.

Model 1 for the three land use variables does not explain a statistically significant amount of the variation. However, the results are in the expected direction, with out-migration in the years before 2003 associated with a lower amount of land in cultivation, with the effects being the strongest for land under intensive cultivation. In general, the effects are strongest for migration that occurs immediately before 2003 and are reduced over time. Where migration had occurred three years before 2003 there were very small differences in the land under cultivation compared to households that did not contain migrants.

In Model 2 the values of the dependent variable measured in 2000 are entered as independent variables. For the total amount of land under cultivation, the effects of migration status are statistically significant after controlling for other predictor variables. Households with migrants cultivate significantly less land than households with no migrants. The effects are most marked in the one and two years immediately after migration. The effects are also very strong, with migrant households cultivating over 20 percent less land than non-migrant households. Compared to Model 1, the migration variable in Model 2 becomes much stronger after controlling for the initial level of land under cultivation. In Model 1 the effect is muted because those households with migrants are also the households with the highest amount of land under cultivation.

Most of the household structure variables have no relationship with the total amount of land under cultivation. However, there are positive relationships with

Table 5.4 Unstandardized ordinary least squares regression coefficients of models predicting proportion of household members working in agriculture

Variable	Proportion of Household members in Agriculture 1	2
No out-migration (ref)		
Out-migrant 1 year before	0.08**	- 0.04*
Out-migrant 2 years before	0.11**	- 0.03*
Out-migrant 3 years before	0.15**	- 0.03*
Number in-migrants 2000	-	0.01
Number in-migrants 2003	-	0.02*
Male out-migrant 15-59 2003	-	0.07**
Female out-migrant 15-59 2003	-	0.14**
Proportion household members in agriculture 2000	-	0.54**
Number HH members 2000	-	0.07**
Number HH members 2003	-	- 0.11**
HH female members in agriculture 2000	-	- 0.10**
HH female members in agriculture 2003	-	0.28 **
Dependency ratio 2000	-	0.00*
Dependency ratio 2003	-	- 0.00**
Household wealth 2000	-	- 0.01*
Household wealth 2003	-	- 0.00
Household debt 2000	-	0.00
Household debt 2003	-	0.00*
Constant	0.48	0.31
N	4511	4511
R-square	0.027**	0.535**

Note: * - significant at .05 level; ** - significant at .01 level.

the household wealth and the total amount of household debt. This suggests that higher levels of household resources are associated with more land used for cultivation. Debt in this instance can be viewed as a form of household resource as it is likely to be used to fund cultivation.

As might be expected, given the dominance of intensive agriculture in overall agriculture, the results for intensive agriculture are very similar to those for the total land under cultivation. Households with migrants cultivate significantly less land in intensive agriculture than households with no migrants.

However, the results for non-intensive agriculture are very different from those for total land under cultivation and land under intensive cultivation. Out-migration has no impact on the amount of land that is used for non-intensive cultivation. This is probably because relatively little labour is required for this form of agriculture.

Household Members in Agriculture

From Table 5.4, the dependent variable is proportion of household members in agriculture in 2003, after controlling for other predictor variables, the relationship between household migration status and the proportion of household members employed in agriculture switched from positive to negative, but it remained significant. Households with one or more members who had migrated before the survey employed a higher proportion of household members in agriculture than households with no migrants. This effect remains even after controlling for the proportion of household members employed in agriculture in the year 2000. The primary reason for the shift in sign was the inclusion of the variables that denoted whether the household had a male or female migrant of labour force age that migrated from the household in the period 2002–2003. The signs for both of these variables were positive and the relationships were statistically significant.

Combined, the results suggest that out-migration contributes to decreased proportion of household members engaged in agriculture, although the net effects are not large. In-migration of either males or females may help free up other household labour to work in agriculture. The positive coefficient for in-migration in 2003 indicates that households that contain in-migrants increase the proportion of labour allocated to agriculture. There are also positive relationships between proportion of household members in agriculture in 2003 with proportion of household members engaged in agriculture in 2000 and household female members engaged in agriculture in 2003.

However, there is a negative relationship between the proportion of household members in agriculture and the number of females engaged in agriculture in 2000. A potential reason for this is that it is likely that migrants are drawn generally from those households which have an excess of females employed in agriculture. A proportion of these migrants are not replaced in agriculture after migration. Where a household has a large number of females in agriculture, migration of females may be encouraged because females are likely to remit earnings to the household and because agricultural work is less preferred for females when other employment options are available.

The effects of other predictor variables on the proportion of household members employed in agriculture in 2004 are mostly not statistically significant. Higher household dependency ratios in 2000 are associated with a lower proportion of household members working in agriculture, which is opposite to the direction of relationship between household dependency ratios in 2004 and the proportion of household members engaged in agriculture. This is probably related to the fact that although the labour force increased in the later year, these people did not engage in agriculture work, partly because they are older, tend to be more highly educated and get jobs in non-agriculture sectors. As expected, households with higher living standards tend to have a low proportion of household members working in agriculture.

Conclusions

There are strong motivations for migration. Opportunities are not distributed evenly over space and migration is one of the few ways through which new opportunities can be accessed. Although there is a consensus that migration does provide economic benefits to migrants and usually to their origin households, the effects of migration on household production are unclear. In this study we address this issue for agricultural households.

The results of the study are mixed. There is some evidence from the multivariate analysis to suggest that out-migration results in a reduction in the amount of land used for agriculture, particularly intensive agriculture. This effect is most marked in the period immediately following migration. However, households with more land in cultivation are also those that are most likely to have migrants. It appears that agricultural households that have more resources, as measured by the amount of land in cultivation are more likely to be able to fund one or more members to migrate. This subsequently reduces the amount of land under intensive cultivation.

However, this effect is relatively short-lived. Households respond to the loss of household members through migration by adjusting the proportion of household members engaged in agriculture. The adjustment is not large, but it is significant. The data does not allow us to determine whether the increase in agricultural labour is accomplished through a reallocation of labour from non-agricultural to agricultural activities or through the employment in agriculture of household members not previously employed. However, it is clear that agricultural households in Kanchanaburi do have the labour resources available to compensate for the loss of labour through migration.

The interplay between migration and household allocation decisions is complex and this chapter has only touched the surface of this relationship. The results of our study suggest that households do face constraints in the amount of land that they can cultivate and the amount of labour they can use in agriculture immediately after the out-migration of household members. However, households soon adjust to these constraints, drawing on existing household resources to substitute for the labour of household members who have migrated. There is no evidence to suggest that out-migration seriously impacts upon the agricultural production of households in this context. Although this chapter does not address the issue of whether the out-migration of household members benefited the household through the receipt of migrant remittances,[2] if this did occur migration from agricultural households is likely to have added to the economic viability of the households.

2 Remittance data is not available for all years and hence we are not able to examine this relationship.

References

Aphichat, C., Archavanitikul, K., Richter, K., Guest, P., Thongthai, V., Boonchalaksi, W., Piriyathamwong, N. and Vong-Ek, P. 1995. *The National Migration Survey of Thailand*, Institute for Population and Social Research Publication No. 188, Bangkok: Mahidol University.

Bilsborrow, R. 1998. The state of the art and overview of the chapters, in *Migration, Urbanization and Development: New Directions and Issues*, edited by R. Bilsborrow, Norwell, Massachusetts: UNFPA and Kluwer Academic Publishers, 1–56.

Francis, E. and Hoddinott, J. 1993. Migration and differentiation in western Kenya. *The Journal of Development Studies*, 30(1), 115–45.

Glytos, N. 1993. Measuring the income effect of migrant remittances: a methodological approach applied to Greece. *Economic Development and Cultural Change*, 42(1), 130–68.

Guest, P. 1998a. Assessing the consequences of internal migration: methodological issues and a case study on Thailand based on longitudinal household survey data, in *Migration, Urbanization and Development: New Directions and Issues*, edited by R. Bilsborrow. Norwell, Massachusetts: UNFPA and Kluwer Academic Publishers, 275–318.

Guest, P.1998b. *The Dynamics of Internal Migration in Viet Nam*. UNDP Discussion Paper 1, Hanoi: UNDP.

Institute for Population and Social Research 2004. *Kanchanaburi Project 2003*. Bangkok: Institute for Population and Social Research.

Joekes, S., Heyzer, N., Oniang'o, R. and Salles, V. 1994. Gender, environment and population. *Development and Change*, 25(1), 137–65.

Makannah, T. 1988. Remittances and rural development in Sierra Leone. *Peasant Studies*, 16(1), 53–62.

Makinwa-Adebusoye, P. 1993. Labour migration and female-headed households, in *Women's Position and Demographic Change*, edited by N. Federichi, K.O. Mason and S. Sogner. Oxford: Claredon Press, 319–38.

Nelson, N. 1992.The women who have left and those who have stayed behind: rural-urban migration in central and western Kenya, in *Gender and Migration in Developing Countries*, edited by S. Chant. London: Belhaven Press, 109–38.

Oberai, A., Prasad, P. and Sardana, M. 1989. *Determinants and Consequences of Internal Migration in India: Studies in Bihar, Kerala and Utar Pradesh*. Delhi: Oxford University Press.

Singhanetra-Renard, A. 1992. The mobilization of labour migrants in Thailand: personal links and facilitating networks, in *International Migration Systems: a Global Approach*, edited by M. Kritz, L.L. Lim and H. Zlotnik. Oxford: Claredon Press, 190–204.

Skeldon, R. 1997. Rural-to-urban migration and its implications for poverty alleviation. *Asia-Pacific Population Journal*, 12(1), 3–16.

Skeldon, R. 2002. Migration and poverty. *Asia-Pacific Population Journal*, 17(4), 67–82.

Stark, O. 1991. *The Migration of Labour*. Cambridge: Basil Blackwell.

United Nations Development Program (UNDP). 2007. Sufficiency Economy and Human Development, in Thailand Human Development Report. New York: UNDP.

Chapter 6
Migration and Socio-Economic Change in Rural South Africa, 2000–2007

Mark A. Collinson, Annette A.M. Gerritsen, Samuel J. Clark,
Kathleen Kahn and Stephen M. Tollman

Introduction

Migration and Socio-Economic Status

As discussed in Chapter 1, migration is fundamentally linked to changes in the socio-economic status (SES) of individuals and households. It is usually seen as a livelihood strategy (Stark and Bloom 1985, Lucas 1997, Quisumbing and McNiven 2007), however not all migrants are successful and the links to socio-economic status in the sending household depend on whether or not the migrant becomes employed (Aliber 2003).

Previous research has shown that a large proportion of migrants in Africa remit back to their rural home. For example Adepoju shows in 1974 that approximately 60 percent of the migrants in Oshugbo (Nigeria) were sending money to their home area (Adepoju 1974), while Johnson and Whitelaw found 89 percent of migrants sending money out of Nairobi (Johnson and Whitelaw 1974). Gubert shows that 51 percent of households in a rural Senegalese population receive remittances from migrants abroad (Gubert 2002). Mechanisms are discussed in the theories such as New Economics of Labour Migration (NELM) (Taylor 1999) and Transnational Migration (Kuhn 2005). The NELM posits that in the presence of imperfect markets or credit constraints migration may complement productivity in rural areas by relaxing credit or risk constraints; relative deprivation may serve as a stimulus or trigger for migration. Theories of transnational migration bring in the importance of social networks to the migration decision and choice of destination. Migrant networks are social ties between household members and previous migrants from the same household, neighbourhood or village. This network provides the social, economic and political solidarity that underpin the flow of information, investment and trade (Portes 1996, Faist 2000). The theory suggests that most migrants go to where they have more connections, not necessarily where they will earn the most (Massey and Espinosa 1997). There is some consensus among authors that the relevant level of analysis is the household or community (Azam and Gubert 2006), since investment is required to send a migrant and improve their education prior to migration.

The motivating force that keeps remittances flowing has been a subject of study. There are ties that bind the migrant to the rural household and two positions are prominent. Altruistic theories argue that migrants act to improve the welfare of family members (Agarwal and Horowitz 2002) and the remittances respond to the needs of families. The other is that there are contractual arrangements, also called 'enlightened self-interest' (Lucas and Stark 1985), in which remittances represent the outcome of an implicit contract between the migrant and the household. Lucas and Stark examine prevailing motives for remittance behaviour in Botswana and conclude that both altruistic and contractual motivations occur (Lucas and Stark 1985). Migrants provide an insurance against hard times, as evidenced by the remittance amount increasing when droughts threaten livestock, thus migrants exhibit altruism (Azam and Gubert 2006). Yet, wealthier families receive more than poorer ones, which they interpret as migrants defending their inheritance, thus self-interest. The same blend of motivations is supported by work in Kenya (Hoddinott 1994). Van Wey explores it in Thailand and finds that both are important but discriminated by gender and socio-economic status. Most women and poorer men behave more altruistically while most men and better-off migrants behave more contractually (Van Wey 2004).

A frequently asked question in the literature is the extent to which migration is positive or negative for the households and communities that send them. There is no one answer and researchers have found support for each perspective. Rempell and Lobdell review 50 studies in developing world settings in a 1976 article and show consumption, education and quality of housing are most likely to be impacted by remittances, but return migration is more important than remittances for local development (Rempel and Lobdell 1978). On reviewing the literature in Thailand, Skeldon shows that the poorest community members tend be left behind by wealthier out-migrants. The impact of remittances are more positive from international migration than internal, but the impact of internal remittances can also be substantial (Skeldon 1997); a finding confirmed by Kuhn in Bangladesh (Kuhn 2005). Return migrants also contribute to communities through bringing back new ideas and attitudes toward family size and education. Skeldon concludes that migration can have negative impacts for sending communities, but the balance is positive. Guest shows that remittances produce income multiplier effect in rural economies and that remittances tend to reduce inequality among rural households (Guest 1998). At the macro level, in developing countries, Chen et al. (1998) report that net rural-to-urban migration is positively correlated with gross national product growth and indicators of SES and health.

The arguments that migration is negative for rural development cover a range of important perspectives. The main issue is that out-migration can exacerbate labour shortages leading to negative net impacts on farm incomes (Lucas 1997, Quisumbing and McNiven 2007). Migrants may earn less than non-migrants with equivalent qualifications in their place of destination, hence comprising a large segment of the urban poor (Guest 2006). Lipton builds this case and argues that out-migration increases inter-household inequality, because only the better off households can benefit from remittances (Lipton 1980). Furthermore, dependence on remittances serves as a means of retaining traditional systems in rural areas and therefore serves as a brake for development (Azam and Gubert 2006). Azam and Gubert also argue that although migrant remittances are used for investing in agricultural production, the overall contribution is negative because households that rely on migrant remittances are less driven to farm efficiently. Van Wey reports that the level of community organization in Mexico influences whether there are benefits to the whole community from migrant remittances. If community structures are organized they can solicit support for community projects from migrant remittances which otherwise would increase inequity (Van Wey et al. 2005).

The debate continues and longitudinal data are needed to help resolve it. Typically, inferences about the impact of migration are made from censuses or broad cross-sectional studies as discussed by White in Chapter 1 of this volume. Without longitudinal data it is hard to disentangle selection from causality or general development from selective improvement.

Poverty in South Africa

Post-apartheid South Africa is arguably one of the world's most socio-economically unequal societies (Leibbrandt et al. 1999, Klasen 2002). Wealth inequality in South Africa occurs both between and within race groups and across geography with particularly large differences between urban and rural areas. The Project for Statistics on Living Standards and Development (PSLSD) led by the Southern Africa Labour and Development Research Unit of the University of Cape Town is a comprehensive national household survey conducted in 1993 that has supported a range of studies on poverty. Key findings include that while inter-race inequality remained a huge challenge there is also significant inequality within race groups, especially within the African population (Leibbrandt et al. 2000). At a national level, rural, former homeland areas are among the poorest communities (Aliber 2003). The poverty rate in rural areas is 73 percent, more than three times the rate in metropolitan areas. Moreover, the poor households in rural areas are much poorer than their urban counterparts. On average in rural areas households need to increase income by over 70 percent to reach the poverty line, while this figure is 40 percent in metropolitan areas (Klasen 1997). In another study, econometric techniques are used to decompose inequality within the rural population by type of income to determine what contribution each income type makes in resolving inequality between households (Leibbrandt et al. 2000). Three income types emerge as most influential: remittances from temporary migrants, income from locally employed household members and government grants. These three income types influence inter-household inequality the most, but do so in different ways. Income from remittances and grants tend to lower inter-household inequality and are most important at the poorer end of the socio-economic spectrum, while incomes from local employment actually increase inter-household inequality and are more relevant in the upper half of the distribution. Other sources of income considered include agricultural and capital income, but these do not play a significant role in the distribution of household socio-economic status.

Other research based on the PSLSD derives classes of poverty within the poor rural population. This work highlights the diversity of livelihoods employed by rural households. Relating the survey data to qualitative data in KwaZulu-Natal it is determined that around half of rural households were poor, but for purposes of intervention, it is necessary to move beyond headcount poverty and isolate livelihood classes (Carter and May 1999). While household assets varied by livelihood class, more fundamental to escaping poverty are differentials in entitlements (Sen 1981) representing the capabilities of households to use assets productively (Carter and May 1999). The poorest class of households have both asset and entitlement poverty.

A number of studies investigate the dynamics of poverty at the household level in order to describe the movement of households between socio-economic levels and identify factors associated with chronic poverty (stagnation) and either improvement or deterioration (movement). Important among these is the KwaZulu-Natal Income Dynamics Study (KIDS) comprised of a panel starting in 1993 with 1,400 households, 1,183 of which were located and interviewed again in 1998 (May et al. 2000; Carter and May 2001). Findings from KIDS reveal that income varies substantially over time for many households, but that there are some households that are stuck and cannot rise out of extreme poverty. These are termed chronically poor in contrast to transitory poor, those that are able to change SES over the course of the study. The poverty dynamics among the rural households observed by the KIDS study show that 22 percent of households are chronically poor, eleven percent move from poverty to non-poor, 19 percent move from non-poor to poor and 47 percent stay non-poor (better-off) throughout (Aliber 2003).

These studies from South Africa make clear that rural households bear a large burden of national poverty with about half of the households above the poverty line, while in the worse-off half there is dynamic movement into and out of poverty with almost a quarter of households remaining chronically poor. Panel studies from a range of other African countries highlight the high proportion of transitory poor household (Hoddinott et al. 2003). Studies examining factors correlated with chronic poverty show that demographic events – death of an adult bread-winner, losing or gaining a job, a fall or rise in migrant remittances or a fall or rise in non-labour income (that is, government grants) – are the major driving factors moving households into or out of poverty. Statistically, getting a job is more important than changes in earnings when it came to lifting a rural household out of poverty (Woolard and Klasen 2004). Finally, the KIDS sample in South Africa shows that aggregate poverty rates increased from 27 percent to 43 percent in the 1993–1998 period (Carter and May 2001).

In 2007 Statistics South Africa published a report that compares poverty statistics between the 1996 national census, 2001 census and 2007 community survey. The report attributes increases in the standard of living between 1996 and 2007 to the success of government economic policies, particularly among the most disadvantaged groups. They conclude that 'substantial progress has been made with regard to improving the living conditions of South Africans' (Statistics South Africa 2007). The results show positive improvements at national and provincial level with the proportion of people with no formal education decreasing from 19 percent in 1996 to ten percent in 2007 and the proportion with some secondary education increasing from 34 percent to 40 percent. Census data from (inland) Mpumalanga Province shows improvements in living conditions in the period 2001–2007. The proportion of households using electricity for lighting increases from 69 percent to 82 percent; use of electricity for cooking increases from 38 percent to 56 percent; access to a cell-phone increases from 32 percent to 73 percent, fridge ownership from 51 percent to 64 percent and television ownership from 54 percent to 66 percent.

Given existing findings it is likely that migration plays a key role in the socio-economic and poverty dynamics of rural communities, but providing evidence to support this assertion is difficult for two reasons. First, migration is a very diverse phenomenon with both permanent and temporary patterns, each with different gender dynamics, underlying motivations and potentially different outcomes. Second, there is an important endogeneity in the relationship between migration and poverty. This can be conceived as a bi-causal relationship such that less-poor households are more likely to send migrants because they can more easily overcome the costs of migration, such as migrant support and child care constraints (Ardington et al. 2007) and conversely, migration can improve household SES through migrant remittances and enhanced social networks (Massey 1990, Guest 2006). This can also be expressed by noting that households are not equally likely to send a migrant, neither are they likely to be equally poor and that the same factors may influence each of these characteristics, hence there is likely to be unobserved heterogeneity. These problems are addressed in this study using longitudinal data that is sensitive to migration type and includes repeated measures of migration and household SES to tease apart selection from causation.

Aims of the Chapter

1. To evaluate the trend in household socio-economic status in a typical former 'Bantustan' rural community of South Africa.
2. To examine patterns and trends in remittance behaviour amongst migrants.
3. To examine the relationship between migration and change in household socio-economic status.
4. To determine which households are chronically poor and how this relates to migration and other household factors.

Data and Methods

The Agincourt Health and Socio-Demographic Information System

The Agincourt Health and Socio-Demographic Surveillance System (HDSS) is a field and computer operation that routinely updates a population register for the entire, contiguous, sub-district of 70,000 people in the northeast of South Africa. An annual update has been made of each birth, death or migration since the baseline in 1992. The principle is to maintain a dynamic list of all people living and who have lived, within the geographically defined sub-district (Tollman 2008). A field operation is conducted each year to visit every household in the sub-district (11,988 households were visited in 2005). Trained and supervised fieldworkers interview the best respondent available who should be knowledgeable about household events. During this interview the fieldworker verifies existing records, records new data pertaining to individuals or the household and systematically records the demographic events that have occurred since the preceding year's census update (Tollman et al. 1999, Kahn et al. 2007, Tollman 2008). This is supplemented by a maternity history of all in-migrant women aged 15–55 years, as well as residence histories and other modules built into the census (see below). A dynamic household roster showing current members is printed onto each census form in advance of the annual update. Individual attributes are recorded at first observation, that is, baseline, in-migration or birth; include name, sex and data of birth, mother's identity, mother status and nationality or refugee status. Relationship to household head is recorded as a time-changing variable. The census update is conducted by four teams of six fieldworkers each with a supervisor who uses GIS[1]-based maps to ensure that every household is covered. The maps are kept up to date by taking GPS[2] readings of new dwellings each year. A verbal autopsy is conducted on each death to establish the cause. The verbal autopsy interview is conducted by a trained lay fieldworker in the vernacular, that is Shangaan and assessed by medical practitioners to establish the main cause of death, as well as immediate and contributing causes (Kahn et al. 1999, 2000, Kahn 2006, Tollman et al. 2008). Thus, a prospective, longitudinal database of demographic events for the entire sub-district population has been established and regularly updated for sixteen years.

Each year special modules are nested within the update round to provide basic information relevant to particular research lines (Kahn 2006). The variables are selected due to their salience in current scientific literature, knowledge of the local population and results of extensive piloting. Most cross-sectional modules are repeated to allow for longitudinal analysis, but with different periodicities depending on the expected pace of change. This chapter employs data from three modules in particular, household assets, labour force participation and temporary

1 Geographic Information System.

2 Global Positioning System.

migration. A cross-sectional household asset survey is conducted every second year, that is 2001, 2003, 2005, in which the key features of assets owned by households are recorded. The questionnaire contains 34 ordinal variables, covering such areas as building materials and structure of the main dwelling, access to water and power supply and ownership of appliances, transport and livestock.

Labour force participation modules were completed in 2000 and 2004. These record key features of labour force participation on all de jure persons in the sub-district aged ten years or older. The definition of 'working' and categories of unemployment were derived by starting with conventional definitions and undertaking a process of discussion and refinement with local field staff and community members. For the study, 'work' was defined as an activity that brought income or resources into the household from outside and 'locally employed' was defined as the employment of a person who is not a temporary migrant.

A temporary migrant is a household member who was physically absent for at least six months in the year preceding the interview but who remained a de jure household member while away. A permanent migrant is someone who either entered or exited the study area (sub-district) permanently, that is, without intent to move back or go anywhere else. A temporary migration census module was conducted in 2002 and 2007. People who were identified as temporary migrants were entered into the module and a household respondent answered questions about the migration. Key areas included the duration of migration, destination, reason for migration, return pattern, communication pattern, remittances, linked moves and child care arrangements. Temporary migration is a state of circulation that can be repeated or terminated and a person can be a temporary migrant for many years as long as they remain a part of the linked household.

Analytic Approach

An asset-based 'absolute SES' indicator was constructed from the household asset surveys. In constructing this indicator the aim was to keep it as simple as possible, retain some sense of a real, additive scale so that it could be compared through time and finally to recognize that assets fall into importantly different broad groups. To begin each asset variable was coded with the same valence (that is, increasing values correspond to greater SES) and effectively given equal weight by rescaling so that all values of a given asset variable fall within the range (0, 1). Assets were then categorized into five broad groups – 'modern assets', 'livestock assets', 'power supply', 'water and sanitation' and 'dwelling structure'. For each household within each asset group, the rescaled asset values were summed and then rescaled again to yield a group-specific value in the range (0, 1). Finally for each household these five group-specific scaled values were summed to yield an overall asset score whose value could theoretically fall in the range (0, 5). In reality, values for the overall 'absolute SES' indicator, range from 0.75–4.0. The final overall score effectively gives equal weight to the five asset groupings and within each group to each of the individual assets. A number of other more complex asset indicators were constructed and compared to one another and to the individual asset values. This indicator is highly correlated with the others and more correlated with the individual asset values and since it is far easier to calculate and explicate it was chosen for our final analysis.

To examine the patterns and trends of remittance data from two cross-sectional temporary migration modules (2002 and 2007) are used. This enables comparison of sex differentials for employed temporary migrants, the trends in type and amount of remittance by sex of migrant and frequencies of migrant destination categories with the likelihoods of remittance from each destination. Significance testing of trends and differentials were made using Stata 9 (Stata Corporation).

To examine SES change in relation to migration and other household factors a household panel is derived from the HDSS by creating a database of household/years. For each household/year combination resident household members are evaluated for the value of key attributes and the data aggregated up to a household level. Some, like absolute SES or receipt of government grants, are intrinsically household level variables. Factors are allowed to change over time and can be measured from a baseline controlling for other key factors associated with change in the outcome variable. Two types of regressions are run using Stata 9 (Stata Corporation). The first is an Ordinary Least Squares (OLS) regression which estimates the impact of migration and household factors on SES at the end of the period (2005). The second is a logistic regression that estimates the impact of migration, both temporary and permanent, other livelihoods and household factors on the likelihood of households not remaining chronically poor for the duration of the study. A chronically-poor household is defined as having an SES level below the median value in 2001 and not crossing over that threshold through 2003 and 2005. Throughout the presentation of the regression results, the valence of the coefficients we present is such that increasing coefficient values share a positive relationship with SES.

Migration and Household Factors

Migration and other household factors are extracted from the main HDSS database in order to classify the experience and exposure of individual people. Migration events include male and female adult temporary and permanent migration. The number of adult temporary migrants of both sexes is computed for 2001, 2002, 2003, 2004 and 2005. Permanent migration is captured as events, either permanent in- or out-migration, during each year of the study. These events are summed into categories for each combination of male and female in and out-migration. For both permanent and temporary migration, migrants are defined as adults aged 15–64 because migration is more likely to be directly linked to changes in SES in this age group compared to other ages.

The relationship between temporary migration and poverty is conceptualized as having two components: selection and causation. The selection component relates to the fact that households with temporary migrants may have been better off to begin with before migration and therefore more likely to send a migrant initially. This selection effect is represented in the model by levels of temporary migration at the start of the period, that is, 2001. The causal component is the effect of temporary migration on poverty net of selection effect and other factors in the model. The causal component is related to change in the number of temporary migrants after the baseline measurement.

Key household variables known to be associated with SES through earlier research are included in the models. These are: household size, gender of the household head and nationality of household head which discriminates South African from Mozambican headship. The latter are former refugee households that escaped civil war in Mozambique in the mid to late 1980s and are now self-settled immigrants in these rural villages. These variables can potentially confound the relationship of migration and SES and is handled in a similar way to temporary migration. The level of livelihood engagement is controlled for at the start and not allowed to change, while change variables are introduced four years later.

Results

Socio-Economic Changes 2001–2005

Figure 6.1 shows the distribution of absolute household SES in three successive rounds of data collection from 2001–2005. The distribution of household SES is a wide-based normal curve and the whole curve shifts slightly upwards (toward higher SES) with each successive round of data collection. The largest gains are made in the centre of the distribution (especially between 2003 and 2005) showing that households near the middle rather than at the tails of the distribution experienced the greatest improvement in SES.

Under apartheid, poverty was exacerbated by the migrant labour system that forced a large proportion of the South African black population to relocate to arid, unproductive rural 'Bantustans'. Scholars have shown that in the 1990s the parts of the South African settlement system with the highest levels of poverty were the former Bantustan areas. Apartheid ensured that the socio-economic level from which to measure change was low. Adding to this has been the impact of the HIV/AIDS pandemic that has been linked to high levels of labour migration and the settlement dynamics of southern Africa (Collinson et al. 2007). The increase in mortality has been accompanied by a reduction of life expectancy of about 12 years in males and females, with most of the excess mortality occurring in the prime age adults (Kahn 2006, Schatz and Ogunmefun 2007). Despite these negative contextual factors this study reports an overall improvement in SES of the study population. This agrees with findings of the national statistics agency (Statistics South Africa 2007) that identify improvement in measures of socio-economic wellbeing in the former homeland areas.

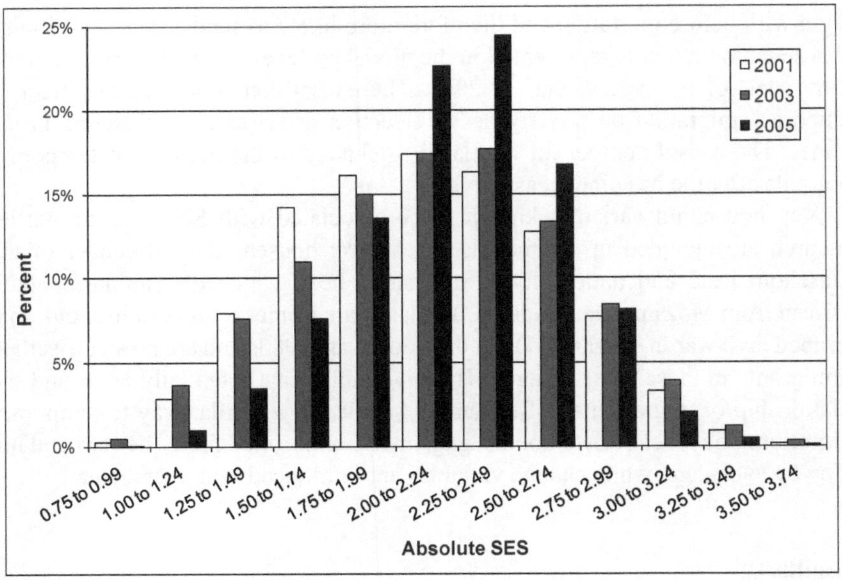

Figure 6.1 The distribution of socio-economic status. Histogram of the percentage of households by socio-economic status showing changes in poverty levels between three successive rounds of data collection from 2001–2005

The absolute SES indicator used in this study incorporates information on a range of assets owned or available to the household and groups them into five groups. Ownership of modern assets like televisions, fridges and stoves shows strong improvement. Power supply shows an improvement due to the expansion of electricity coverage and water and sanitation and quality of housing show some improvement. Communications improve vastly with cell phone ownership expanding such that even poor families invested in cell phones to keep in touch with family members and absent temporary migrants. Livestock ownership is the only category of assets to decline consistently over the period.

The analysis conducted in this study investigates two aspects of the dynamics of socio-economic change: change in SES across the whole SES distribution and whether a household remains below the SES baseline median for the duration of the study, that is, is 'chronically poor'. Factors associated with these SES dynamics are examined in multi-variate regressions below. First, we examine remittance behaviour of temporary migrants by sex and destination category.

Remittance Behaviour 2002–2007

Data on temporary migrant remittances are available in 2002 and 2007. The trends and sex differential is examined. Among adult migrants of both sexes males are significantly more likely to remit back to the rural household (53 percent of male migrants and 42 percent of female migrants in 2007; p-value 0.000). Some adult migrants are employed others are not, but if we consider only the migrants that are employed, females are significantly more likely to remit (63 percent of employed male migrants and 67 percent of employed female migrants; p-value 0.002). These statistics are examined in 2002 and 2007, but there is no change over time in the percent of temporary migrants remitting. However, for the percent of employed temporary migrants remitting there is a significant downward trend: from 66 percent to 63 percent in employed male temporary migrants and from 70 percent to 67 percent in employed female temporary migrants.

We now examine what is remitted. Table 6.1 shows that it is not only money and that there is a significant sex differential in the type of remittance. Male migrants are more likely to remit money alone (77 percent versus 60 percent in 2007; p-value 0.000), while female migrants are more likely to remit money and food. Female migrants are also more likely to remit only clothes or only food. Thus female remittances are more diverse than that of males. There is a significant trend in the diversification of remittances from 2002–2007, with money being combined more with food and clothes for both sexes of migrants.

Table 6.2 shows the amount of money remitted by sex and year. It shows a significant gender differential. Female migrants are much more likely to remit lower amounts and male migrants larger amounts with the cross over at the mode, that is, R301–500. In 2002, the mean amount for all remittances is R445 and the median is R350, standard deviation R382, with a highly skewed distribution comprising a long right tail and a skewness coefficient of 7.2. Over the period of observation the amounts remitted monthly show a significant increase. In 2007, the median is R500, mean R622, standard deviation R728, with an even skewer distribution, a very long right tail and a skewness coefficient of 11.3. Thus, the range of monthly remittance amounts is increasing over time.

The destinations of temporary migrants are presented in Table 6.3. There is a range of destinations of varying distances. In most migrant destinations there is a significant gender differential. Female temporary migrants are more likely to move shorter distances and to nearby commercial farms and game farms (12 percent of males go to farms and 17 percent of females; p-value 0.000). Secondary urban places are more likely to attract male than female temporary migrants (30 percent of males and 27 percent of females; p-value 0.000). The most popular destination for males and females is the primary metropolis of Johannesburg/Pretoria which is the national capital and industrial centre (47 percent of male migrants and 46 percent of female migrants; p-value 0.044). Although the percent by gender going to the main metropolis are similar they are statistically different with a higher proportion of males. The proportion of migrants that remit varies significantly by

Table 6.1 Type of remittance by sex and year, with p-values showing the sex differential in each year; and the trend 2002–2007 for each sex

Type of remittance	Male migrants in 2002	%	Female migrants in 2002	%	Male v. female 2002; p-value	Male migrant in 2007	%	Female migrants in 2007	%	Male v. female 2007; p-value	Male 2002 v.2007; p-value	Female 2002 v.2007; p-value
Money only	3,376	86	990	66	0.000	3,471	77	1,054	60	0.000	0.000	0.001
Clothes or food only	307	8	304	20	0.000	254	6	227	13	0.000	0.000	0.000
Money and clothes or food	226	6	198	13	0.000	420	9	299	17	0.000	0.000	0.002
Money, clothes, food	10	0	8	1	0.111	334	7	169	10	0.004	0.000	0.000
Total	3,919	100	1,500	100		4,479	100	1,749	100			

Table 6.2 **Average monthly amount remitted by sex and year, with p-values showing the sex differential**

Monthly amount	Male migrants in 2002	%	Female migrants in 2002	%	Male v. Female 2002; p-value	Male migrants in 2007	%	Female migrants in 2007	%	Male v. Female 2007; p-value
<R200	797	21	551	38	0.000	447	11	244	16	0.000
R201–300	810	21	441	30	0.000	472	12	304	20	0.000
R301–500	1,185	31	357	24	0.000	1,263	32	559	37	0.001
R501–700	447	12	60	4	0.000	596	15	181	12	0.003
R701–1,000	474	12	42	3	0.000	847	22	161	11	0.000
>R1,000	148	4	14	1	0.000	276	7	45	3	0.000
Total	3,861	100	1,465	100		3,901	100	1,494	100	

Table 6.3 Destinations of migrants by sex, with p-values showing the sex differential; and the percent of migrants of both sexes remitting from each destination category

Destination category	Male migrants at the destination category in 2007	%	Female migrants at the destination category in 2007	%	Male v. Female 2007; p-value	% migrants (both sexes) remitting from that destination
Nearby village or town	302	3	302	7	0.000	51%
Same province agriculture/ game farm	1,063	12	726	17	0.000	76%
Same province secondary urban	2,667	30	1,190	27	0.000	55%
Other province	549	6	145	3	0.000	55%
Primary metropolis	4,208	47	1,999	46	0.044	38%
Other country	23	0.3	8	0.2	0.388	53%
Unknown	69	0.8	21	0.5	0.048	27%
Total	8,881	100	4,391	100		49%

destination category chi^2 (6) = 892, P = 0.000). Seventy six percent of migrants on commercial farms and game farms remitted, while only 38 percent of migrants in the primary metropolis remitted. This is explained by the fact that farms are most likely to be employment destinations whereas migrants have other reasons to go to the metropolis, such as for education or visiting family members.

Migration and Socio-Economic Change

This section describes the relationship between migration and poverty dynamics. Table 6.4 displays results from a multivariate regression model used to investigate migration and other household factors associated with changes in socio-economic status. In addition to migration two other livelihood strategies are included that are hypothesized to be important in affecting SES: local employment and the receipt of government grants. The causal effects of temporary migration on poverty can be seen in the second and fourth rows of the tables (for males and females respectively) as coefficients on the 'change' variables.

Temporary Migration Selection Effects

A significant selection effect exists between migration and household SES. This arises from the fact that generally better-off households are more likely to send migrants and thereby also more likely to receive remittances. Consequently within a generally poor population in which migration is a key livelihood strategy, it is the comparatively better-off households that both send migrants and receive remittances. Additionally, temporary migration builds migrant social networks that are then available to the household. These networks in turn make it easier for the household to send additional temporary migrants. In this way a positive reinforcement feedback system is created around temporary migration that significantly encourages additional and continued temporary migration from households that can make the initial investment. This in turn adds to the endogeneity of the relationship between migration and SES.

The proportion of households with one or more male temporary migrant was high and stable, increasing only slightly from 49 percent to 51 percent between 2001 and 2005. Throughout this period male temporary migration and household SES share a positive association. Table 6.4 is an OLS regression on SES at the end of the period by migration and other household factors. Factors are displayed as either selection or causation effects. The selection coefficient (row one of Table 6.4) for households with one male temporary migrant at the beginning of the period is positive and significant: 0.04 (0.03, 0.05) and the coefficient for more than one male temporary migrant at the beginning of the period is 0.03 (0.01, 0.05), implying an underlying positive relationship with SES across the distribution of the number of male temporary migrants. These firm, positive results support the hypothesis that having male temporary migrants in the household is positively associated with SES.

For female temporary migration, the selection effects are similar to those observed for males. These are displayed as the 'level' at start of period variables in row three of Table 6.4. Having one female temporary migrant in the household is positively associated with SES. There is a highly significant positive coefficient of 0.03 (0.02, 0.05) for one female temporary migrant at the beginning of the period and a significant positive coefficient of 0.03 (0.0, 0.7) for more than one female temporary migrant at the beginning of the period. The confidence intervals are wider because there are fewer cases of households with this level of female temporary migration. In sum, having female temporary migrants in the household is positively associated with SES.

This positive relationship between male and female temporary migration and SES that we observe here can be explained as a positive selection of a subgroup of households that are both better off and more likely to send a migrant. This selection effect of male and female adult temporary migration shows an endogenous relationship between male and female temporary migration and SES.

Causal Effects of Temporary Migration

This section investigates the causal effects of temporary migration over time net of the selection effect. The question 'does temporary migration improve SES' is addressed by including 'change' variables in the multivariate regression model. These include both positive and negative changes, increases and decreases in the number of temporary migrants sent by a household or permanent in and out-migrations or death of a temporary migrant. The death of a prime-age temporary migrant is major shock to overall household income and this population has been heavily impacted by HIV mortality that often affects temporary migrants (Hunter et al. 2007, Schatz and Ogunmefun 2007). If there is a casual relationship between temporary migration and poverty, then reducing the number of temporary migrants associated with a household will negatively affect SES. The estimated coefficients associated with the causal or 'change-effect' variables are displayed in second and fourth (for males and females respectively) rows of Tables 6.4.

Looking first at the potential positive effects, male temporary migrants increased in 34 percent of households. Much of this increase is the result of increases in temporary migration among young men. There is a modest but significant increase in the odds of improving SES as a result of increasing male temporary migration, 0.01 (0, 0.03), comparing households with a positive change in the number of male temporary migrants to households with no change. The effects of increasing the number of female temporary migrants associated with a household, 0.03 (0.01, 0.04) show a slightly stronger positive effect.

Next we examine the effects of decreasing the number of temporary migrants. Decreasing the number of temporary migrants of either sex leads to statistically significant decreases in SES. There is a sex differential in this finding with the loss of male temporary migrants having a more negative effect on SES. Across the period 2001–2005, a decrease in the number of male temporary migrants led to a change in absolute SES of -0.04 (-0.06, -0.02). For females these numbers are: -0.03 (-0.05, -0.01).

In summary, the scale of male temporary migration is double that for females although both streams have grown stronger over the study period 2001–2005. Temporary migration has a significant causal impact on SES over this period for male and female migration. There is a strong negative impact of reducing the number of temporary migrants associated with a household and this is slightly worse when a household loses a male temporary migrant.

Other Livelihoods

Two other livelihood strategies that are shown to be important in earlier research are described by our data: local employment and receipt of government. Data on these livelihoods are available in 2000 and 2004. Data on local employment in 2000 and 2004 shows that the proportion of households with locally employed members dropped from 37 percent in 2000 to 34 percent in 2004. This variable is strongly associated with SES. In the regression model in Table 6.4 it is possible to investigate associations at baseline in 2000 and changes in local employment over the period to 2004. The baseline shows a positive association with SES, but both positive and negative changes in the number of locally employed household members are also important and highly significant. Increasing the number of locally employed household members has a positive impact on SES, 0.03 (0.01, 0.04). Decreasing the number of locally employed household members reduces household SES with a coefficient of -0.05 (-0.07, -0.03). Table 6.4 shows that 25 percent of households experienced an increase in the numbers of locally employed members after baseline and 20 percent experiencing a decrease and that these changes had very substantial effects on SES.

The number of grants received by households in the study population increases dramatically over the period of observation. In 2000, 16 percent of households are receiving a grant whereas in 2004, 40 percent are receiving a grant and 18 percent receive more than one grant. The impact of grants on poverty changes significantly over the period. Estimates for levels of grant receipt at the start of the period are neither important nor significant, mainly due to low levels of grant receipt. Over the period it is possible to investigate the dynamics of grant receipt net of other factors in the model. The existing level of grant receipt in 2000 was associated with poverty (being in the lower half of the SES distribution) because the grants were selectively given to poor households.

Turning to changes in the number of grants received during the period, an increase in the number of grants received, net of the baseline and other change factors, is associated with a highly significant and important improvement in absolute SES during the period, coefficient 0.04 (0.03, 0.05). Only three percent of households receiving a grant at the beginning of the period suffered a decrease in the number of grants they receive, so interpreting the results associated with decreases in the number of grants is dubious. However even without statistical power, the direction and magnitude of the changes in SES are consistent with the fact that grants have a very important affect on SES; decreases in the number of grants are associated with a 0.03 decline in absolute SES.

Household Factors

While isolating the effects of migration on household SES through time we have controlled for other household characteristics that are likely to affect household SES: size of the household, gender of the head of the household and nationality of the head of the household. These 'control' variables are indeed strongly associated with household SES and it is worth considering the estimates associated with them.

Household size has an important and highly significant relationship with absolute SES. In 2001, medium-size households with four to eight members comprise 51 percent of all households and have a substantially higher SES than smaller households with one to three members, associated with an estimated

Table 6.4 OLS regression on household socio-economic status at the end of the period, by migration and other household factors from 2001–2005

Type of factor	Factor	Level	n	%	Coefficient: SES absolute
Baseline	SES level at start				0.45 (0.44, 0.47)***
Selection	Male adult temporary migration in HH at start of period	no male temporary migrants	5,675	51	ref
		1 male temporary migrant	3,945	36	0.04 (0.03, 0.05)***
		>1 temporary migrants	1,507	13	0.03 (0.01, 0.05)**
Causation	Change in male adult temporary migration in HH	no change	5,911	53	ref
		increase in # male temp migrants	3,753	34	0.01 (0, 0.03)*
		decrease in # male temp migrants	1,463	13	-0.04 (-0.06, -0.02)***
Selection	Female adult temporary migration in HH at start of period	no female temporary migrants	8,265	74	ref
		1 female temporary migrant	2,218	20	0.03 (0.02, 0.05)***
		>1 female temporary migrants	644	6	0.03 (0, 0.07)*
Causation	Change in female adult temporary migration in HH	no change	6,785	59	ref
		increase in # female temp migrants	1,835	16	0.03 (0.01, 0.04)***
		decrease in # female temp migrants	2,920	25	-0.03 (-0.05, -0.01)***
Selection	Number of local income earners in the household at start of period	no local incomes	7,438	63	ref
		1 local income	3,117	27	0.05 (0.03, 0.07)***
		>1 local income	1,199	10	0.08 (0.05, 0.1)***
Causation	Change in number of local income earners in household	no change in number of local earners	6,494	55	ref
		Increase number of local earners	2,867	25	0.03 (0.01, 0.04)***
		decrease number of local earners	2,393	20	-0.05 (-0.07, -0.03)***

Selection	Number of government grants in household at start of period	no government grants	9,513	81	ref
		1 government grants	1,866	16	-0.01 (-0.02, 0.01)
		>1 government grant	375	3	0 (-0.03, 0.03)
Causation	Change in number of government grants	no change in number of grants	5,164	44	ref
		Increase number of grants	6,258	53	0.04 (0.03, 0.05)***
		Decrease number of grants	332	3	-0.03 (-0.06, 0.01)
Selection	Household size at start of period	1-3 members	3132	27	ref
		4-8 members	6017	51	0.05 (0.03, 0.07)***
		9+ members	2637	22	0.06 (0.03, 0.08)***
Selection	Gender of HH head at start of period	Female	3823	34	ref
		Male	7318	66	0.03 (0.02, 0.05)***
Selection	Nationality of HH Head at start of period	Mozambican	3093	28	ref
		South African	8048	72	0.13 (0.11, 0.14)***
		Constant			1.11 (1.08, 1.14)***

Source: Collinson, M.A., Clark, K., Gerritsen, A.A.M. and Tollman, S.H. The dynamics of poverty and migration in a rural South African community, 2001–2005, submitted to *Demography* – under review.

coefficient of 0.05 (0.03, 0.07). Larger households with nine or more members make up 22 percent of all households and also have higher SES than smaller households, estimated coefficient of 0.06 (0.03, 0.08). Clearly larger households tend to have more assets, even after controlling for key livelihood strategies of household members such as migration, local employment and grants transfers.

In 2001, male-headed households make up 66 percent of all households and have higher SES than female-headed households, coefficient of 0.03 (0.02, 0.05). This suggests that female-headed households are at greater risk of being in poverty than male-headed households. Two important reasons are that female headship implies that a prime-age male is not part of the household and therefore there are fewer potential income earners. Additionally, the labour market for women from this poor population is largely informal and pays less on average than the more formal labour market to which males have greater access.

Net of everything presented above, the 28 percent of all households headed by Mozambicans have an extreme deficit in SES compared to South African households. Households headed by South Africans are much more likely to improve their SES over time with a coefficient 0.13 (0.11, 0.14) compared to the

Mozambican households. Despite efforts to make services available and improve the legal status of self-settled Mozambican immigrants, these households are still at a comparative disadvantage with respect to the already disadvantaged South African households in the area (Hargreaves et al. 2004).

Chronically Poor Households

Chronically poor households are those with a level of absolute SES that over the entire period 2001–2005 remain below the median of the distribution of household SES at baseline (2001). Households were categorized according to this definition and the resulting binary variable was regressed against the household factors described above. The odds ratios of a household not being chronically poor are presented in Table 6.5.

Despite a general upward movement of the distribution of household SES there was a chronically poor group comprising 29 percent of households that were unable to lift themselves out of extreme poverty, that is, stayed at the bottom of the distribution and even worse, stayed below the median of the distribution at the beginning of the study. These households represent a trapped subgroup that is immobile at the very bottom of the SES distribution. Consequently, this group is vital to identify in order to design and correctly target poverty alleviation policies.

Temporary migration is related to chronic poverty in important and statistically significant ways, net of the other factors. Temporary migration for both males and females is endogenous with chronic poverty but inversely related such that households with male and female temporary migrants are highly unlikely to be chronically poor. The odds ratios are 1.87 (1.64, 2.12) for one male temporary migrant, 1.95 (1.61, 2.38) for more than one temporary migrant; and 1.5 (1.27, 1.77) for one female migrant and 1.75 (1.32, 2.33) for more than one female temporary migrant in the household. Net of positive selection there is also a causal impact of temporary migration on chronic poverty. Increasing temporary migration of either gender significantly lowers the odds of a household being chronically poor, odds ratio 1.29 (1.14, 1.47) for males and 1.35 (1.17, 1.56) for females. Consistent with this finding, a decrease in temporary migration of either gender significantly heightened the odds of being chronically poor, odds ratio 0.74 (0.62, 0.87) for reduction in the number of male temporary migrants and similar odds of 0.72 (0.58, 0.89) for reduction in the number of female temporary migrants. Clearly temporary migration by household members of either sex is an important household livelihood strategy that protects households from languishing in the chronic poverty. Since male migration is twice as likely to occur as female migration and the odds of escaping chronic poverty are slightly higher for female temporary migration it can be inferred that female migration plays a larger role than male migration from chronically poor households.

The relation of permanent migration to chronic poverty is also examined in Table 6.5. Permanent migration generally does not differentiate chronic poverty, although permanent out-migration of more than one female household member is significantly associated with chronic poverty 0.7 (0.51 – 0.95).

The model then identifies further livelihoods and household factors differentiating the chronically poor over and above the effects described above. Households with more than one household member employed locally at the start of the period are significantly protected from chronic poverty. Increasing the number of locally employed household members significantly improves the odds of not being chronically poor, whereas a reduction in the number of locally employed household members significantly increases the odds of being chronically poor. Households with more than one government grant at the start of the period are significantly protected from chronic poverty. Increasing the number of government grants significantly improves the odds of getting out of chronic poverty, whereas decreasing the number of grants received, even though it only occurs in three percent of households, shows a significant association with chronic poverty.

Households of medium and larger size are significantly protected, as are male headed households. The largest coefficient in the model is the nationality of household head; compared to self-settled Mozambican refugees, South African households are much less likely to be chronically poor ratio 4.78 (4.28–5.34).

Discussion and Conclusion

This chapter examines the processes that relate migration and household SES to each other over time which are complex and endogenous making them very difficult to measure and evaluate. The overall logic of the analytical strategy is to utilize the prospective, longitudinal, whole-population nature of the data in an attempt to unravel the endogenous relationship between migration and household SES. This is done by defining the levels of each indicator at the beginning of the period in order to assess and control for the selection effects that allow households with greater SES to be more able to engage in sending temporary migrants; these result in identification of households at different SES 'levels'. Changes in indicator values at the household level are then assessed over the study period. The level or selection variables can now be used to control for where households start from and isolate the actual effects of changes in various indicators through time, in particular the number of temporary migrants associated with the household. This allows us to investigate changes in household SES and relate those back to SES at the end of the period. The model fits are relatively good with the OLS regression model in Table 6.4 having an R^2 value of 0.44 and the logistic regression model in Table 6.5 having a pseudo R^2 value of 0.13. A factor that somewhat compromises the models' fit is the use of levels that included zero and changes that include decrease, since the zero category cannot decrease.

The SES index used in this study is a summary indicator built from ownership of household assets which enables an examination of the whole SES distribution. Governing our choice of this absolute rather than relative SES indicator is the

Table 6.5 Logistic regression on households that are not chronically poor from 2001–2005, by migration and other household factors

Type of factor	Factor	Level	N	%	Odds ratio (95%CI)
Selection	Male adult temporary migration in HH at start of period	no male temporary migrants	5,675	51	Ref
		1 male temporary migrant	3,945	35	1.87 (1.64–2.12)***
		>1 temporary migrants	1,507	14	1.95 (1.61–2.38)***
Causation	Change in male adult temporary migration in HH	no change	5,911	53	Ref
		increase in # male temp migrants	3,753	34	1.29 (1.14–1.47)***
		decrease in # male temp migrants	1,463	13	0.74 (0.62–0.87)***
Causation	Male permanent out-migration from HH	no male out-migrants	9,948	92	Ref
		1 male out-migrant	735	7	0.94 (0.76–1.16)
		>1 male out-migrants	120	1	1.14 (0.66–1.96)
Causation	Male permanent in-migration into HH	no male in-migrants	10,530	97	Ref
		1 male in-migrant	253	2	1.06 (0.75–1.49)
Selection	Female adult temporary migration in HH at start of period	no female temporary migrants	8,265	74	Ref
		1 female temporary migrant	2,218	20	1.5 (1.27–1.77)***
		>1 female temporary migrants	644	6	1.75 (1.32–2.33)***
Causation	Change in female adult temporary migration in HH	no change	6,785	59	Ref
		increase in # female temp migrants	1,835	16	1.35 (1.17–1.56)***
		decrease in # female temp migrants	2,920	25	0.72 (0.58–0.89)***

Causation	Female permanent out-migration from HH	no female out-migrants	9,011	83	Ref
		1 female out-migrant	1,465	14	0.94 (0.8–1.09)
		>1 female out-migrants	327	3	0.7 (0.51–0.95)**
Causation	Female permanent in-migration into HH	no female in-migrants	10,055	93	Ref
		1 female in-migrant	672	6	1.06 (0.85–1.32)
		>1 female in-migrants	76	1	0.59 (0.33–1.04)*
Selection	Number of local income earners in the HH at start of period	no local income	7,438	63	Ref
		1 local income	3,117	27	1.73 (1.49–2.02)***
		>1 local income	1,199	10	2.64 (2.09–3.32)***
Causation	Change in number of local income earners in HH	no change	6,494	55	Ref
		increase	2,867	24	1.28 (1.11–1.47)***
		decrease	2,393	20	0.63 (0.53–0.76)***
Selection	Number of government grants in HH at start of period	no government grants	9,513	81	Ref
		1 government grants	1,866	16	1.07 (0.93–1.23)
		>1 government grant	375	3	1.95 (1.39–2.72)***
Causation	Change in number of government grants	no change	5,164	44	Ref
		increase	6,258	53	1.53 (1.38–1.7)***
		decrease	332	3	0.74 (0.55–1.01)*
Selection	HH size at start of period	1-3 members	3,132	27	Ref
		4-8 members	6,017	51	1.74 (1.53–1.97)***
		9+ members	2,637	22	2.16 (1.81–2.57)***
Selection	Gender of HH head at start of period	female	3,823	34	Ref
		male	7,318	66	1.49 (1.32–1.67)***
Selection	Nationality of HH head at start of period	Mozambican	3,093	28	Ref
		South African	8,048	72	4.78 (4.28–5.34)***

Source: Collinson, M.A., Clark, S.J., Gerritsen, A.A.M., Kahn, K. and Tollman, S.M. The dynamics of poverty and migration in a rural South African community, 2001–2005, submitted to *Demography* – under review.

fact that investigating SES-migration relationships over time requires nuanced measures of migration behaviour and SES that are comparable through time, so that real rather than relative changes can be measured.

Migration is a key livelihood strategy. It is mostly concerned with household selection, such that households in the better-off half of this generally poor population are more likely to send a male temporary migrant and remain better-off. But there is also a kind of migrant that has to work harder to migrate. These are more driven members of poorer households who are striving to send home income and bring back ideas from more modern places. The chapter explores the hypothesis that temporary migrants have both a causal and an endogenous relationship with SES in the sending community. Remittances from migrants make a difference to the distribution of resources at a community and household level. For the poorest households the most important factors improving SES are government grants and female temporary migrants, while for less poor households male temporary migrants and local employment are most important. There are a higher proportion of men employed and they also remit more, but among migrants who work, the proportion of women who remit is higher than men. Female remittances are particularly important in the poorer end of the SES distribution. The range of remittance types varies with males more likely to send money alone and women to combine it with food and/or clothes and the diversity of remittance types is increasing over time. This is consistent with the increasing rates of unemployment in the population. Nevertheless, despite more diversification the monetary amount of remittance is increasing over time. The data also challenges the stereotype that temporary migrants move mainly to the large cities and remit back to the rural areas. This does occur and the large cities are the most likely destination, but not the most reliable source of remittances. Nearby commercial and game farms have a much higher likelihood of the migrant remitting. These destinations are also differentially more important for female migrants.

The findings on remittances and local employment show that both rural and urban labour markets play critical roles in the welfare of rural households. Policies that improve access to employment for both genders in both rural and urban settings are likely to positively affect rural livelihoods. Government grants primarily affect households in the poorer half of the SES distribution although they also generate some improvement in the less poor half as well. Subgroups that are at heightened risk of being chronically poor are Mozambican refugees, female-headed households and small-sized households. Grants need to reach these marginalized subgroups.

The focus on chronic poverty also emphasizes the importance of temporary migration as a livelihood strategy; a select group exists that sends migrants and possess assets and they are highly unlikely to be chronically poor. Additionally, sending a temporary migrant lowers the risk of being chronically poor, especially in poorer households that send a female temporary migrant. Households that have lost temporary migrants of either gender, through retrenchment, death or permanent out-migration, have a significantly higher risk of being chronically poor.

Labour migration remains a prominent livelihood strategy for household income and asset generation. Temporary migration should be facilitated for men and women and made safer by measures such as enhancing long-distance public transport, improving roads and introducing migrant-oriented health services. Mechanisms for transferring remittances safely back to remote areas are essential. Improving living conditions for the urban poor and farm labourers will benefit the rural poor and should be part of a strategy for national economic development. Migrants from rural areas are positively selected sojourners who access urban labour markets, commercial agriculture and game farms and remit back to rural homes, a fundamental process of generating and distributing resources.

Acknowledgements

The funders who have invested in the Agincourt Health and Socio-Demographic Surveillance System are warmly acknowledged for their vision and making the work possible: Wellcome Trust, UK; Medical Research Council, South Africa; National Research Foundation, South Africa; Department of Science and Technology, South Africa; Anglo-American Chairman's Fund, South Africa; National Institutes of Health and Aging, USA; William and Flora Hewlett Foundation, USA; and Andrew W. Mellon Foundation, USA. The communities of Bushbuckridge are gratefully acknowledged as active and not uncritical partners and the Mpumalanga Provincial Department of Health and Social Welfare who are also vital partners and stake-holders.

References

Adepoju, A. 1974. Rural-urban socio-economic links: the example of migrants in South-West Nigeria, in *Modern Migrations in Western Africa*, edited by S. Amin. London: Oxford University Press, 127–37.

Agarwal, R. and Horowitz, A.W. 2002. Are international remittances altruism or insurance? Evidence from Guyana using multiple-migrant households. *World Development*, 30(11), 2033–44.

Aliber, M. 2003. Chronic poverty in South Africa: incidence, causes and policies. *World Development*, 31(3), 473–90.

Ardington, C., Case, A. and Hosegood, V. 2007. *Labour Supply Responses to Large Social Transfers: Longitudinal Evidence From South Africa*. NBER Working Paper Series 13442, National bureau of economic research.

Azam, J.P. and Gubert, F. 2006. Migrants' remittances and the household in Africa: a review of evidence. *Journal of African Economies*, 15(Suppl 2.), 426–62.

Carter, M. and May, J. 1999. Poverty, livelihood and class in rural South Africa. *World Development*, 27(1), 1–20.

Carter, M. and May, J. 2001. One kind of freedom: poverty dynamics in post-apartheid South Africa. *World Development*, 29(12), 1987–2006.

Chen, N., Valente, P. and Zlotnik, H. 1998. What do we know about recent trends in urbanization? in *Migration, Urbanization and Development: New Directions and Issues*, edited by R. E. Bilsborrow. New York: United Nations Population Fund, 59–88.

Collinson, M.A., Tollman, S.M. and Kahn, K. 2007. Migration, settlement change and health in post apartheid South Africa: triangulating Agincourt demographic surveillance with national census data. *Scandinavian Journal of Public Health*, 35(Suppl. 69), 77–84.

Faist, T. 2000. *The Volume and Dynamics of International Migration and Transnational Social Spaces*. New York: Clarendon Press.

Gubert, F. 2002. Do migrants insure those who stay behind? Evidence from the Kayes area (western Mali). *Oxford Development Studies*, 30(3), 267–87.

Guest, P. 1998. Assessing the consequences of internal migration: methodological issues and a case study on Thailand based on longitudinal household survey data, in *Migration, Urbanisation and Development: New Directions and Issues,* edited by R. Bilsborrow. New York: UNFPA and Kluwer Academic Publishers.

Guest, P. 2006. Bridging the gap: internal migration in Asia, in *African On The Move: Migration In Comparative Perspective*, edited by M. Tienda, S.E. Findley, S.M. Tollman, E. Preston-Whyte. Johannesburg: Wits University Press, 194–216.

Hargreaves, J., Collinson, M., Kahn, K., Clark S. and Tollman S. 2004. Childhood mortality among former Mozambican refugees and their hosts in rural South Africa. *International Journal of Epidemiology*, 33(6), 1–8.

Hoddinott, J. 1994. A model of migration and remittances applied to western Kenya. *Oxford Economic Papers*, 46(3), 459–76.

Hoddinott, J., Quisumbing, A., de Janvry, A. and Woldehanna, T. 2003. *Pathways from poverty: evaluating long-term strategies to reduce poverty*. Basis Brief Number 30. Department of agricultural and applied economics. Madison, USA: University of Wisconsin.

Hunter, L.M., Twine, W. and Patterson, L. 2007. Locusts are now our beef': adult mortality and household dietary use of local environmental resources in rural South Africa. *Scandinavian Journal of Public Health*, 35(Suppl. 69), 165–74.

Johnson, G.E. and Whitelaw, W.E. 1974. Urban-rural income transfers in Kenya: an estimated-remittances function. *Economic Development and Cultural Change*, 22(3), 473–79.

Kahn, K. 2006. *Dying to Make a Fresh Start: Mortality and Health Transition in a New South Africa*. PhD Thesis. Umeå, Sweden: Umeå University

Kahn, K., Tollman, S., Garenne, M. and Gear, J. 1999. Who dies from what? Determining cause of death in South Africa's rural north-east. *Tropical Medicine and International Health*, 4(6), 433–41.

Kahn, K., Tollman, S.M., Garenne, M. and Gear, J.S.S. 2000. Validation and application of verbal autopsies in a rural area of South Africa. *Tropical Medicine and International Health*, 5(11), 824–31.

Kahn K., Tollman, S.M., Collinson, M.A., Clark, S.J., Twine, R., Clark, B.D., Shabangu, M., Gómez-Olivé, F.X., Mokoena, O. and Garenne, M. 2007. Research into health, population and social transitions in rural South Africa: data and methods of the Agincourt health and demographic surveillance system. *Scandinavian Journal of Public Health*, 35(Suppl. 69), 8–20.

Klasen, S. 1997. Poverty, inequality and deprivation in South Africa: an analysis of the 1993 SALDRU survey. *Social Indicators Research*, 41(1–3), 51–94.

Klasen, S. 2002. Social, economic and environmental limits for the newly enfranchised in South Africa. *Economic Development and Cultural Change*, 50(3), 607–42.

Kuhn, R. 2005. A longitudinal analysis of health and mortality in a migrant-sending region of Bangladesh, in *Migration and Health in Asia*, edited by S. Jatrana, M. Toyota and B.S.A. Yeoh. London: Routledge, 177–208.

Leibbrandt, M., Bhorat, H. and Woolard, I. 1999. *Understanding Contemporary Household Inequality in South Africa*. Development Policy Research Unit Working Paper Number 99/25. Cape Town: University of Cape Town.

Leibbrandt, M., Woolard, C. and Woolard, I. 2000. The contribution of income components to income inequality in the rural former homelands of South Africa: a decomposable gini analysis. *Journal of African Economies*, 9(1), 79–99.

Lipton, M. 1980. Migration from rural areas of poor countries: the impact on rural productivity and income distribution. *World Development*, 8(1), 1–24.

Lucas, R.E.B. 1997. Internal migration in developing countries, in *Handbook of Population and Family Dynamics*, edited by M.R. Rosenzweig and O. Stark, Amsterdam: Elsevier Science Publishers: 72–98.

Lucas, R.E.B. and Stark, O. 1985. Motivations to remit: evidence from Botswana. *Journal of Political Economy*, 93(5), 901–18.

Massey, D.S. 1990. Social structure, household strategies and the cumulative causation of migration. *Population Index*, 56(1), 3–26.

Massey, D.S. and Espinosa, K. 1997. What's driving Mexico-US migration: a theoretical, empirical and policy analysis. *American Journal of Sociology*, 102, 939–99.

May, J., Carter, M.R., Haddad, L. and Maluccio, J.A. 2000. Kwazulu-Natal income dynamics study (KIDS) 1993–1998: a longitudinal household data set for South African policy analysis. *Development Southern Africa*, 17(4), 567–81.

Portes, A. 1996. Transnational communities: their emergence and significance in the contemporary world-system, in *Latin America in the World Economy*, edited by R.P. Korzeniewicz and W.V.C. Smith. London: Greenwood Press.

Quisumbing, A.R. and McNiven, S. 2007. *Moving Forward, Looking Back: The Impact of Migrant's Remittances on Assets, Consumption and Credit*

Constraints in Sending Communities in The Rural Philippines. ESA - working paper no. 07-05. The food and agriculture organization of the United Nations.

Rempel, H. and Lobdell, R.A. 1978. The role of urban-to-rural remittances in rural development. *Journal of Development Studies*, 14(3), 324–41.

Schatz, E. and Ogunmefun, C. 2007. Caring and contributing: the role of older women in multi-generational households in the HIV/AIDS era. *World Development*, 35, 1390–403.

Sen, A. 1981. Ingredients of famine analysis: availability and entitlements. *The Quarterly Journal of Economics*, 96(3), 433–64.

Skeldon, R. 1997. Rural-to-urban migration and its implication for poverty alleviation. *Asia-Pacific Population Journal*, 12(1), 3–16.

Stark, O. and Bloom D.E. 1985. The new economics of labour migration. *American Economic Review*, 75(2), 845–67.

Stata Corporation Intercooled STATA 9.0, 4905 Lakeway drive, College station, Texas 77845 USA.

Statistics South Africa 2007. *Community Survey 2007* (Revised version), Statistical release P0301. Pretoria. Available at: www.statssa.gov.za [accessed: 15 June 2008].

Taylor, J. 1999. The new economics of labour migration and the role of remittances in the migration process. *International Migration*, 37(1), 63–86.

Tollman, S.M. 2008. *Closing the Gap: Applying Health and Socio-Demographic Surveillance to Complex Health Transitions in South and Sub-Saharan Africa*. PhD Thesis. Umeå, Sweden: Umeå University.

Tollman, S.M., Herbst, K., Garenne, M., Gear, J.S.S. and Kahn, K. 1999. The Agincourt demographic and health study – site description, baseline findings and implications. *South African Medical Journal*, 89(8), 858–64.

Tollman, S.M., Kahn, K., Sartorius, B., Collinson, M.A. Clark, S.J. and Garenne, S.M. 2008. Implications of mortality transition for primary health care in rural South Africa: a population-based surveillance study. *The Lancet*, 372(9642), 893–901.

Van Wey, L.K. 2004. Altruistic and contractual remittances between male and female migrants and households in rural Thailand. *Demography*, 41(4), 739–56.

Van Wey, L.K., Tucker, C.M. and McConnell, E.D. 2005. Community organization, migration and remittances in Oaxaca. *Latin American Research Review*, 40(1), 83–107.

Woolard, I. and Klasen, S. 2004. *Determinants of Income Mobility and Household Poverty Dynamics in South Africa*. Institute for the Study of Labour (IZA) discussion paper number 1030. Bonn, Germany.

Chapter 7
Parents' Migration and Children's Education in Matlab, Bangladesh

Nurul Alam and Peter Kim Streatfield

Background

Bangladesh is known to the world for its high population density and proneness to natural disasters. Its population is growing at the rate of 1.52 percent annually and one-third of the population is under- or unemployed (Bangladesh Bureau of Statistics 2003). Its economy is agrarian and is close to the limits of availability of agricultural land. Agricultural workforce has remained numerically stagnant for more than a decade (Bangladesh Bureau of Statistics 2005). High under- and unemployment rates provoke migration as part of coping strategies. Migration facilitates removal of excess labour that cannot be absorbed in agriculture and out-migration may help reduce pressure on the land. Natural disasters, high population growth, availability of low per capita land and high unemployment rates force rural people to migrate to other rural areas of low population density, towns, cities or overseas in search of employment or earning opportunities (Afsar 2000). Lein (2003) reported a positive association between proneness to natural hazards (flood and river erosion) and internal migration in Bangladesh.

Migration is a livelihood strategy. Men migrate leaving behind their aging parents, spouses and children to earn livelihood. Migration contributes positively to material well-being, but parents, spouses and children endure lengthy separations that causes grief and loss of attachment. Men undergo a parallel experience and their grief often turns into sadness, guilt and anxiety. With the advent of information and communication technologies, phone contact, especially through mobile phones, is now pervasive and can reduce some of the negative effects of separation. Migration through assimilation, diffusion and remittances affects positively utilization of health services in Guatemala and children's immunization in Ethiopia (Lindstorm and Franco 2006, Kiros and White 2004). A retrospective survey conducted in Matlab, Bangladesh in 1996 showed that father's overseas migration affected positively children's schooling (Khun 2006). Major flow of migration is the rural-to-urban migration; its effect on children's education is unknown.

Rural women in Bangladesh are mostly housewives. They migrate leaving children and other family members behind under adverse conditions such as marital breakdown and economic hardship. Women of broken marriage are vulnerable socially and economically. Some of them migrate to undertake a job as garment factory worker or domestic help in a city or overseas. Their migration leaving children and family members behind is a proxy for poverty rather than empowerment and it may eventually have adverse effects on children's education.

Population-based migration data collected prospectively in Matlab, Bangladesh, provide an opportunity to study the effects of parents' migration by destination on children's education in 2005. Other important factors affecting children's education are age and sex of the child, parents' education and household economic condition (Razzaque and Streatfield 2007, Khun 2006). Lack of control of these factors will confound the effects of parents' migration on children's education.

The objectives of the study are to examine the effects of parents' migration by destination (rural, urban, overseas) on children's school enrolment rate and retention rates (measured with passing of grades five, seven and ten among school enrolled children), controlling simultaneously for a number of demographic and socioeconomic factors; age and sex of the child, parents' education and household asset quintile in rural Bangladesh. The aim is to guide policy makers to undertake remedial actions to the second Millennium Development Goal, 'universal primary education by 2015'.

Methods and Data

This study used population-based migration data recorded along with vital events by the Health and Demographic Surveillance System (HDSS) in Matlab – a rural area of Bangladesh during 1996–2004. Matlab is located 50 km south-west of the capital city, Dhaka. Important occupations of men are agriculture, small business, day labour, service in public, private and NGO sectors and fishing. Women usually do household chores. Young men and women migrate to urban areas to pursue higher education, skill development training and seeking off-farm jobs (Alam and Khuda 2005). They are also attracted to the urban lifestyle.

HDSS residents were followed during 1996–2004 to determine their migration status. An out-migrant is defined as when a HDSS resident in a household who is listed in the last HDSS population census or has become a resident by birth or immigration, has moved out of the HDSS area for at least six months or permanently. An in-migrant is an individual who came into the HDSS after baseline enumeration and is not born there. Migration was recorded by visiting households monthly during 1996–1999 and bi-monthly since 2000. Members of the out-migrant households or their neighbours (in the case when a whole family migrated out) were asked about destination where migrants had gone to and in-migrants were asked the destination they had returned from. Destination of migration (in and out) was categorized into rural, urban, overseas (mainly oil-rich middle-east countries and Malaysia) or none (for no migration).

Table 7.1 Distribution of children (n=63,208) aged 6–19 years by parents' migration destination during 1996–2004

Place of out-migration of fathers	Percent (%)	Place of out-migration of mothers	Percent (%)
None	58.9	None	86.0
Rural	0.9	Rural	2.0
Urban	3.0	Urban	4.2
Long migration*	30.7	Long migration*	7.0
Abroad	5.0	Abroad	0.1
Died	1.5	Died	0.6

Note: *migrated out before 1996 and the exact destination was unknown.

HDSS conducted a household socioeconomic census in 2005, to record individuals' education (type of school attended and grade completed) and household possessions of durable articles (land, cot, wardrobe, chair and table, quilt, watch, radio, television and bicycle). Household durable articles were used to calculate household asset quintile using principal components analysis (Gwatkin et al. 2000). The higher the asset quintile the better was the household economic condition.

Children's education was linked to parents' education and migration status and household asset quintiles. The fathers of some children had been living outside the HDSS area since before 1996 for job and business reasons as had some mothers mostly due to marital breakdown. They are categorized as long-term out-migrants and their education was categorized as 'unknown'. Unlike mothers, fathers visit family occasionally.

Data Analysis

This study analysed education of children aged 6–19 years (as on 1 January 2005) at the start of the school year for a number of independent variables. The dependent variables were:

1. Enrolled in school among children aged six to nine years,
2. Given they were enrolled in school, passed grade five among children aged 12–15 years,
3. Passed grade seven among children aged 14–17 and
4. Passed grade ten among children aged 16–19 years.

The independent variables were age and sex of the child, parents' education, parents' out-migration by type of place during 1996–2004 and household asset quintile. Data were analysed using bivariate and logistic regression techniques. Bivariate relationships are shown in the form of enrolment rates and grade passing rates and multivariate relationships are expressed as odds ratios. Logistic regression was used to estimate odds ratios for each variable (adjusting for clustering of children

Table 7.2 Demographic and socioeconomic factors associated with children's school enrolment and passing grade five, Matlab HDSS area 2005

Demographic and socioeconomic factors	% enrolled in school (and no. of children)	Adjusted odds ratio[a] (and 95% CI)	% passed grade 5 (and no. of children)	Adjusted odds ratio[a] (& 95% CI)
Sex of the child:				
Male	87.2 (10,169)	1	59.4 (9,138)	1
Female	89.7 (9,722)	1.32** (1.21 – 1.45)	68.1 (9,536)	1.57** (1.46 – 1.68)
Child's age (in yr):				
6 (12 for grade 5)	79.3 (5,319)	1	43.5 (4,682)	1
7(13 for grade 5)	88.5 (4,877)	2.20** (1.96 – 2.47)	60.3 (4,465)	2.26** (2.06 – 2.48)
8 (14 for grade 5)	91.9 (4,701)	3.33** (2.94 – 3.78)	71.9 (4,942)	4.08** (3.72 – 4.48)
9 (15 for grade 5)	94.7 (4,994)	5.40** (4.69 – 6.22)	79.2 (4,585)	6.47** (5.86 – 7.14)
Father's education:				
None	83.4 (7,233)	1	53.7 (7,016)	1
Primary (I-V)	89.4 (4,357)	1.41** (1.24 – 1.61)	62.7 (4,119)	1.34** (1.22 – 1.47)
Secondary (VI+)	94.7 (3,134)	1.80** (1.49 – 2.17)	80.9 (3,005)	2.14** (1.89 – 2.42)
Unknown	90.7 (5,167)	1.39 (0.96 – 2.01)	69.1 (4,534)	1.33* (1.12 – 1.58)
Maternal education:				
None	81.7 (7,843)	1	51.6 (8,976)	1
Primary (I-V)	91.0 (7,070)	1.93** (1.73 – 2.16)	69.5 (6,538)	1.83** (1.69 – 1.98)
Secondary (VI+)	96.4 (4,368)	4.18** (3.46 – 5.05)	89.7 (2,661)	6.18** (5.34 – 7.16)
Unknown	88.0 (610)	2.24 (1.41 – 3.52)	71.3 (499)	1.92** (1.41 – 2.62)
Father's migration:				
Did not migrate	87.3 (11,848)	1	61.6 (11,354)	1
Rural-urban	81.6 (239)	0.91 (0.61 – 1.35)	53.9 (141)	0.85 (0.55 – 1.33)
Urban-rural	87.4 (899)	1.02 (0.79 – 1.32)	58.4 (514)	0.99 (0.79 – 1.23)
Migrated long ago	90.3 (5,441)	1.01 (0 71 – 1.44)	68.2 (5,434)	1.20* (1.02 – 1.41)
Migrated abroad	93.8 (1,315)	1.55** (1.20 – 2.00)	71.1 (942)	1.43** (1.21–1.69)
Father died	77.8 (149)	0.63* (0.41 – 1.00)	56.7 (289)	1.04 (0.79–1.36)
Mother's migration:				
Did not migrate	88.6 (16,901)	1	63.9 (16,662)	1
Rural-urban	85.7 (685)	0.79 (0.61 – 1.03)	64.9 (282)	1.14 (0.86 – 1.52)
Urban-rural	89.3 (1,399)	1.14 (0.91 – 1.42)	60.8 (664)	0.95 (0.77 – 1.16)
Migrated long ago	85.6 (825)	0.55** (0.39 – 0.79)	65.6 (956)	0.89 (0.72 – 1.10)
Mother died	83.9 (81)	0.86 (0.46–1.62)	47.3 (110)	0.71 (0.48–1.05)
HH asset quintile:				
Lowest	78.9 (4,368)	1	43.2 (3,219)	1
Second	87.2 (4,005)	1.59** (1 40 – 1.81)	55.4 (3,329)	1.50** (1.34 – 1.67)
Middle	90.6 (4,049)	1.99** (1.73 – 2.28)	63.8 (3,810)	1.87** (1.67 – 2.08)
Fourth	92.7 (4,079)	2.14** (1.83 – 2.49)	72.8 (4,193)	2.43** (2.17 – 2.71)
Highest	94.4 (3,393)	2.35** (1.95 – 2.83)	77.5 (4,123)	2.83** (2.52 – 3.18)
All	88.4 (19,891)		63.8 (18,674)	

Note: Dependent variable was coded 1 if a child enrolled in school or passed grade five, otherwise it was coded 0. *P<0.01, **P<0.001 (all odds ratios are adjusted for clustering of children to the same household).

to the same household, using Stata version eight), controlling simultaneously for the effects of all other variables.

Results

The 2005 household socioeconomic census enumerated 63,208 children (31,203 were boys and 32,005 were girls) aged between six and 19 years. As expected migration was more frequent for fathers than for mothers of children (39.6 percent versus 13.4 percent) during 1996–2004 (Table 7.1). The rural-to-urban migration was more common than the rural-to-rural migration. Long-term migration and overseas migration was more frequent among fathers than among mothers. In the follow-up period death was found more frequent among fathers than among mothers. Some of the differences could be due to age difference; men are older than their wives by 7.5 years in the study population (ICDDR,B 2007).

Among children aged six to nine years, 88.4 percent were enrolled in schools and the enrolment rate (in percentage) was higher for girls than for boys by 2.5 percent (Table 7.2). The school enrolment rate increased gradually with increasing age. Parents' education was associated with increased enrolment rates, with children benefiting more from the education of mothers than from that of fathers. The rate was higher if fathers did migrate abroad than if fathers did not migrate at all. It was lower among children whose fathers died than those whose fathers did not migrate and survived. It was also lower if mothers died (but the difference was not statistically significant). The long-term migration of mothers (mostly due to marital breakdown) was related to lower school enrolment rate compared with mothers' non-migration status. School enrolment rate increased gradually with increase in household asset quintile.

Among school enrolled children who were aged 12–15 years, 63.8 percent passed grade five and the pass rate (in percentage) was higher for girls than for boys by 8.7 percent. The pass rate increased with increasing age and parents' education. The rate of increase was higher for increase in mother's education than for increase in father's education. The long-term and overseas migration of fathers were associated with higher pass rates. That was not the case for the migration of mothers. Household asset quintile was positively associated with pass rate as well.

About half (49.3 percent) of the school enrolled children aged 14–17 years passed grade seven and the pass rate increased with increasing age (Table 7.3). The odds ratio of passing grade seven was higher for girls than for boys and higher for children whose parents have some education than for children whose parents have no education. Compared to the fathers' education, the odds ratios were larger for the same mothers' education. Children whose fathers migrated for a long period or migrated abroad, passed at higher rates than children whose fathers did not migrate during 1996–2004. The chances of passing grade seven were higher among children whose fathers died than among children whose fathers did

Table 7.3 Demographic and socioeconomic factors associated with children's passing grade seven and grade ten, Matlab HDSS area 2005

Demographic and socioeconomic factors	% passed grade 7 (children aged 14-17)	Adjusted odds ratio (95% CI)	% passed grade 10 (children aged 16-19)	Adjusted odds ratio (95% CI)
Sex of the child:				
Male	43.7 (8,859)	1	13.8 (7,934)	1
Female	54.4 (9,748)	1.72** (1.61–1.84)	11.6 (9,313)	0.74** (0.67–0.82)
Child's age (in yr):				
14 (16 for grade 10)	34.2 (4,942	1	7.7 (4,698	1
15 (17 for grade 10)	46.5 (4,585)	1.96** (1.78–2.15)	11.0 (4,382)	1.67** (1.43–1.94)
16 (18 for grade 10)	57.7 (4,698)	3.23** (2.94–3.54)	15.8 (4,163)	2.61** (2.26–3.01)
17 (19 for grade 10)	60.2 (4,382)	3.72** (3.36–4.07)	16.9 (4,004)	2.88** (2.50–3.33)
Father's education:				
None	37.1 (6,750)	1	6.2 (5,889)	1
Primary (I-V)	45.7 (3,986)	1.31** (1.18–1.44)	8.4 (3,442)	1.41* (1.16–1.70)
Secondary (VI+)	69.4 (3,113)	2.11** (1.88–2.37)	24.4 (2,536)	2.68** (2.24–3.20)
Unknown	56.3 (4,758)	1.35** (1.16–1.58)	16.3 (5,080)	1.71** (1.35–2.17)
Maternal education:				
None	33.7 (9,026)	1	5.3 (8,089)	1
Primary (I-V)	57.0 (6,441)	2.07** (1.92–2.24)	13.7 (5,640)	1.82** (1.84–2.38)
Secondary (VI+)	84.3 (2,294)	7.21** (6.29–8.28)	40.4 (1,911)	6.95** (5.95–8.10)
Unknown	61.6 (846)	1.59** (1.33–2.24)	12.8 (1,607)	1.43 (0.97–2.11)
Father's migration:				
Did not migrate	46.1 (11,125)	1	11.0 (9,699)	1
Rural to rural area	48.6 (109)	0.99 (0.61–1.61)	15.0 (93)	1.52 (0.72–3.20)
Rural to urban area	44.8 (395)	1.06 (0.82–1.37)	12.1 (273)	1.50 (0.96–2.37)
Migrated long ago	54.3 (5,853)	1.24* (1.07–1.44)	15.2 (6,219)	1.54** (1.19–1.98)
Migrated abroad	62.0 (755)	1.79** (1.5–2.13)	13.4 (568)	1.32 (0.98–1.77)
Father died	44.6 (370)	1.34* (1.05–1.70)	9.6 (395)	1.68** (1.14–2.50)
Mother's migration:				
Did not migrate	48.9 (16,420)	1	12.7 (14,565)	1
Rural to rural	57.4 (202)	1.61** (1.14–2.63)	19.1 (141)	1.34 (0.79–2.27)
Rural to urban	49.5 (491)	1.00 (0.80–1.25)	14.6 (328)	1.05 (0.69–1.61)
Migrated long ago	53.1 (1,377)	0.93 (0.75–1.14)	11.8 (2,110)	1.02 (0.72–1.45)
Mother died	43.6 (117)	1.40 (0.93–2.11)	6.8 (103)	1.06 (0.49–2.29)
HH asset quintile:				
Lowest	23.0 (2,816)	1	2.7 (2,208)	1
Second	35.9 (3,212)	1.67** (1.47–1.89)	6.0 (2,837)	1.92** (1.41–2.63)
Middle	46.4 (3,677)	2.29** (2.03–2.59)	10.4 (3,341)	2.78** (2.08–3.72)
Fourth	59.3 (4,302)	3.34** (2.95–3.77)	15.8 (4,178)	3.77** (2.85–4.99)
Highest	67.7 (4,600)	4.49** (3.97–5.08)	20.0 (4,683)	4.50** (3.40–5.95)
All	49.3 (18,607)	not applicable	12.6 (17,247)	not applicable

Note: Dependent variable was coded 1 if a child enrolled in school or passed grade five, otherwise it was coded 0; *P<0.01, **P<0.001 (all odds ratios are adjusted for clustering of children to the same household).

Table 7.4 Effects of parents' duration of out-migration (in months) on children's education, Matlab HDSS area 2005

Parents out-migration	Enrol in school Odds ratio[a] (95% CI)	Pass grade 5 Odds ratio[a] (95% CI)	Pass grade 7 Odds ratio[a] (95% CI)	Pass grade 10 Odds ratio[a] (95% CI))
Father	1.002 (0.999-1.005)	1.002 (1.000-1.003)	1.002 (1.001-1.003)	1.003 (1.001-1.005)
Mother	0.997 (0.994-1.000)	0.999 (0.997-1.001)	0.999 (0.998-1.002)	1.001 (0.998-1.004)

Note: a – Adjusted for age, sex, parents' education and household wealth index.

not migrate and survived. The pass rate was higher if mothers migrated to other rural areas compared to those whose mothers did not migrate out. The higher the household asset quintile the higher was the pass rate.

Among children aged 16–19 years enrolled in school, 12.6 percent passed grade ten and the pass rate increased with increasing age. In contrast to other age groups, the pass rate was lower for girls than for boys for the higher rate of discontinuation in girls after grade seven. The pass rate was higher for children whose parents have some education compared with children whose parents do not have education. It was also higher if the education of the father was unknown, suggesting that they may be a selective group. The education of mothers appeared to have a larger effect than that of fathers on children's education. Compared to those whose fathers did not migrate during 1996–2004, those whose fathers migrated for a long-time or overseas migration had higher pass rates. The chances of passing grade ten were higher among children whose fathers died than among children whose fathers did not migrate and survived. This is contrary to our expectation and needs a qualitative study in order to understand the association. The migration of mothers did not have education effects. The higher the household asset quintile the higher was the rate of passing grade ten.

The duration of migration of fathers and mothers was estimated excluding children whose parents died during 1996–2004. For children whose parents migrated before 1996, duration was assigned the longest follow-up period – 104 months. The duration of the migration of parents, particularly the out-migration of fathers was found significantly and positively associated with children's education (Table 7.4).

Table 7.5 shows that children whose fathers did migrate for long periods or did migrate abroad were more likely to belong to top household asset quintiles (fourth or fifth), than children whose fathers did not migrate at all or migrated to other rural areas or urban areas. As expected, children whose mothers did migrate (except for long-term) were not more likely to belong to the top quintiles.

Table 7.5 Parents' migration status and percent of children aged 6–19 years who belong to top household asset quintiles (4th or 5th)

Father's migration status	% in top two quintiles	Mother's migration status	% in top two quintiles
Did not migrate	39.9	Did not migrate	42.7
Rural to rural area	34.9	Rural to rural area	40.3
Rural to urban area	35.9	Rural to urban area	35.4
Migrated long ago	46.2	Migrated long ago	48.0
Migrated abroad	63.7	Migrated abroad	42.7
Died	36.6	Died	35.1
All	42.8	All	42.8

Discussion

The main advantage of this study over other migration studies is that it is prospective in recording out-migration (including destination) by asking household members or neighbours. In rural Bangladesh, men migrated mostly for economic reasons (78 percent of cases) and women leaving children behind migrated mostly in adverse social and economic conditions (Alam and Khuda 2005). Our results show that the duration of migration of fathers, not of mothers was associated with higher chances of passing grades five, seven and ten. Not always was the migration of fathers beneficial to the child's education. Only their long-time migration and overseas migration were associated with increased children's school enrolment and retention rates. Migration may have contributed indirectly to children's education by reducing household income poverty as it was found positively associated with household asset quintile. These effects are in addition to the findings that urban migration experience and having relatives abroad were associated with greater use of formally trained health providers and higher rate of immunization of children in Guatemala and Ethiopia (Lindstorm and Franco 2006, Kiros and White 2004).

Similar to another study (Khun 2006) age of children was found to be positively associated with higher rates of school enrolment and retention in grades five, seven and ten. These associations suggest that a significant proportion of children enrolled in school pass specific grades after modal age. Consistent with the earlier study results (Razzaque and Streatfield 2007, Khun 2006), children of educated mothers and fathers and those of well-off households were more likely to be enrolled in school and continue education compared to their counterparts. The odds ratios of school enrolment and passing grades five, seven and ten reveal that mothers' education was more strongly associated with children's education than fathers' education. The odds ratios were much lower among children of fathers and mothers who had no education than among children of fathers and mothers who had secondary or higher level education. The former discontinued education at higher levels because they were unable to pass exams.

Our results show that the effects of parents' education and household economic conditions are very large. High discontinuation among pupils of poor or uneducated parents suggests that they need additional care to enable them to perform well in school and keep them in school. In-depth studies may help understanding their needs to perform well.

The hidden objectives of the government female secondary school stipend programme introduced in 1994 for rural girls were to eliminate the gender difference, keep the girls in school and thereby to delay marriage and child bearing age. The results show that the stipend programme eventually eliminated the gender gap in grade seven and below. Girls are statistically less likely than boys to complete grade ten and it is true for all girls irrespective of their household economic condition. This reveals that the stipend programme did not achieve its objectives. One of the possible reasons for high discontinuation is the pressure for early marriage of the girls (Mahmud 2003). Parental incentives to early marriage are increasing vulnerability of adolescent girls to sexually aggressive behaviour on the way to and from school (Alam et al. 2009) and the demand for dowry.

Policy Relevance

Migration of men, particularly children's fathers, resulted in a small, but significant improvement in children's education. The primary determinants of children's education are parents' education and household economic condition. This reveals that government planning and adequate budgetary allocation for universal primary education and secondary education for girls are not enough to achieve 'universal primary education' by 2015 (MDG-2). The school stipend programmes (primary and secondary) need to be reviewed so that the children of poor parents with no education can harvest the benefits of the stipends. Household economic development is needed to bring sustained improvement in children's education. Migration of fathers may have contributed indirectly to children's education by improving household economic condition. The sustained absence of parents may have some negative repercussions, which were not studied, on caring of aged parents and other vulnerable household members.

Conclusions

The long-term and overseas migration of fathers increased children's education. Parental education and household economic condition have much larger effects on children's education than father's migration.

Acknowledgements

This study used the data of the Matlab Health and Demographic Surveillance System supported by DFID and ICDDR, B: Centre for Health and Population Research. This research was supported by ICDDR, B. ICDDR, B gratefully acknowledges these donors who provide unrestricted support to the Centre's research efforts: Australian International Development Agency (AusAID), Government of Bangladesh, CIDA, The Kingdom of Saudi Arabia (KSA), Government of the Netherlands, Government of Sri Lanka, Swedish International Development Cooperative Agency (SIDA), Swiss Development Cooperation (SDC) and Department for International Development, UK (DFID).

References

Afsar, R. 2000. *Internal Migration and the Development Nexus: The Case of Bangladesh.* Bangladesh Institute of Development Studies. Dhaka, Bangladesh: University Press Limited.

Alam, N. and Khuda, B. 2005. *Rural-urban Migration in Bangladesh: Causes and Consequences.* Population association of Pakistan: Sharing Population and Development Research across South and West Asia, Fifth Annual Research Conference Proceedings, Karachi, 14–16 December, 2004.

Alam, N., Roy, S.K. and Tahmeed, A. 2009. Sexually harassing behaviour against girls in rural Bangladesh: implication for achieving MDGs. *Journal of Interpersonal Violence.* 19 May.

Bangladesh Bureau of Statistics 2003. *Bangladesh Population Census 2001 Report (provisional).* Planning Division, Ministry of Planning. Bangladesh Bureau of Statistics.

Bangladesh Bureau of Statistics 2005. *Statistical Pocketbook of Bangladesh 2005.* Planning Division, Ministry of Planning. Bangladesh Bureau of Statistics.

Gwatkin, D.R., Rutstein, S., Johnson, K., Pande, P.R. and Wagstaff, A. 2000. *Socioeconomic Differences in Health, Nutrition and Poverty 2000.* HNP/ Poverty Thematic group of the World Bank. Washington, D.C.: The World Bank.

ICDDR,B. 2007 – *ICDDR,B Health and Demographic Surveillance System-Matlab: 2005 Socioeconomic Census*, Vol. 38. Dhaka-1212, Matlab: ICDDR,B.

Kiros, G. and White, M.J. 2004. Migration, community context and child immunization in Ethiopia. *Social Science and Medicine,* 59(12), 2603–16.

Kuhn, R. 2006. The effect of father's and siblings' migration on children's pace of schooling in rural Bangladesh. *Asian Population Studies*, 2(1), 69–92.

Lein, H. 2003. *Hazards and Forced Migration in Bangladesh.* Department of Geography. NTNU Norwegian University of Science and Technology.

Lindstorm, D.P. and Munoz-Franco, E. 2006. Migration and maternal health services utilization in rural Guatemala. *Social Science and Medicine*, 63(3), 706–21.

Mahmud, S. 2003. *Female Secondary School Stipend Programme in Bangladesh: A Critical Assessment.* Dhaka, Bangladesh: Bangladesh Institute of Development Studies (BIDS).

Razzaque, A. and Streatfield, P.K. 2007. Family size and children's education in Matlab, *Bangladesh. Journal of Biosocial Science*, 39(2), 245–56.

PART III
Migration and Health

Chapter 8
Assessing the Effect of Mother's Migration on Childhood Mortality in the Informal Settlements of Nairobi

Adama Konseiga, Eliya M. Zulu, Philippe Bocquier, Kanyiva Muindi, Donatien Beguy and Yazoumé Yé

Introduction

Due to globalization, huge economic inequities and civil strife, millions of people are on the move; some from rural to urban areas and vice versa and others across national borders in search of better economic opportunities or fleeing from disorder and danger at home. Health professionals are concerned about the health status of migrants because people are potential agents for the transmission of infectious diseases and disease spreading agents from the place of origin to the destination. Secondly migrants are likely to be vulnerable to various infections if they relocate to destinations where they are exposed to diseases that were not existent or not common in the place of origin (Bogin 1988, Prothero 1977). The vulnerability of migrants may also be exacerbated soon after moving due to lack of information about health services or preventive mechanisms in the new place of residence.

Migration is a very important determinant of urban population growth in sub-Saharan Africa. Indeed, despite the fact that urban areas in the region have considerably lower fertility rates than rural areas, urban populations are growing at much higher rates relative to rural populations because of the disproportionately high proportion of urban residents who are in the reproductive age range and the net effect of rural-urban migration.[1] A study carried out among Nairobi residents showed that the proportion of Nairobi city-born residents is at most 20 percent for men and women of varying age groups and that half of the residents came to Nairobi between 17 and 23 years old (Agwanda et al. 2008). The majority of slum residents were born in rural areas, with about 70 percent and 91 percent residents of Korogocho and Viwandani slum settlements aged 12 years and above being born in rural areas (Zulu et al. 2009). The failure of urban economies to generate enough jobs and of local authorities to provide adequate housing, basic amenities and other social services for the rapidly growing urban population have forced many poor urbanites to live in slum settlements

1 Apart from natural increase and migration, urban populations also increase due to expansion of urban boundaries and promotion of new areas into urban centres.

because they cannot afford rent elsewhere[2]. According to estimates by UN-Habitat, about 72 percent of all urban residents in sub-Saharan Africa are estimated to live in slum settlements since they lack the basic amenities associated with planned urban residence (UN-Habitat 2003).

Slums are sanctuaries for poor health because they are characterized by poor access to clean water, proper sanitation, garbage disposal and drainage system, overcrowding, poor housing conditions and excessive environmental and air pollution from factories. In fact, the growth of slums and the associated poor health outcomes in these settlements have been touted as the primary reason for the decline in the extent of advantage that urban areas have traditionally had over rural areas in various health outcomes in sub-Saharan Africa (Gould 1998). Slum dwellers generally exhibit higher levels of morbidity, indulgence in risky sexual behaviours and drug abuse, lower utilization of health services and higher mortality than other population subgroups, including rural residents (Zulu et al. 2002, Magadi et al. 2003, Dodoo et al. 2007, APHRC 2002a, Mugisha and Zulu 2004, Ndugwa and Zulu 2008). The biggest inequities in health outcomes between slum and non-slum populations are observed among children (APHRC 2002a). For instance, data from the slums of Nairobi show that children living in slum settlements are considerably more likely to get sick from infectious diseases, less likely to use medical services and more likely to die than other major sub-populations, including rural residents (APHRC 2002a, Kyobutungi et al. 2008). While rural infant mortality and child mortality rates for rural areas were 76 and 113 in 1998, the equivalent rates for slum settlements were 91 and 151, respectively.

A number of studies have examined the effect of migration on child health in general urban populations in sub-Saharan Africa. For example, Brockerhoff's analyses of cross-sectional Demographic and Health Survey (DHS) data have shown that children of rural-urban migrants experience a much higher risk of under-five mortality than children of urban non-migrants, even after the mother has lived in the city for many years. On the other hand, children of rural-urban migrants exhibit higher survival probabilities than children of rural natives (Brockerhoff 1990, see also, Bocquier et al. Forthcoming). Brockerhoff (1995) also found that the disadvantage faced by the migrants is more pronounced in big cities than in smaller urban centres and that the excess mortality of migrant children in big cities is concentrated in low-quality housing areas. A study in Ethiopia found that children of rural-urban migrant mothers had lower immunization coverage than those of non-migrant mothers (Kiros and White 2004). The study attributed this difference to the poor social networks and poor integration into the new place of residence for the migrants. These differences remained significant after controlling for the migrant mother's duration of residence, suggesting that being a migrant had an independent and long-lasting effect on health outcomes for children. A study

2 Many people choose also to live in the relatively cheap squatter settlements in order to accumulate savings for various investments in their home communities while acquiring the city experience that prepares them for a more permanent formal urban job.

based on Indian data, however, showed that the relationship between rural-urban migration and child mortality was not statistically significant after controlling for socioeconomic and health utilization factors (Stephenson et al. 2003). These results show that the effect of migration on health outcomes may vary depending on the setting and nature of migration.

Health indicators for slum dwellers are likely to shape national health indicators and the capacity of countries to achieve the health-related millennium development goals. Given the central role of migration in shaping urbanization patterns in Africa and the uniquely poor health outcomes exhibited by slum dwellers, it is important to understand how their health status is affected by migration. Despite this need, the literature on migration and urbanization is deficient of studies on migration dynamics among slum dwellers, let alone the implications of these dynamics on health outcomes. In this study, we contribute to the understanding of the effect of migration on child health by using longitudinal data to compare infant mortality rates for children born in slum settlements to mothers who are recent migrants compared to long-term residents. Since all the children that we are studying were born in the slums and over the same period, the analysis allows us to determine whether having a mother who is new in the slum setting increases or reduces the risk of child mortality after controlling for various confounding factors.

Study Population and Methods

Study Site and Population

The study utilizes data from the Nairobi Urban Health and Demographic Surveillance System (NUHDSS), implemented by the African Population and Health Research Centre (APHRC). The NUHDSS covers two slum settlements, Korogocho and Viwandani, in Nairobi City, Kenya. The NUHDSS involves visits to all households in the study areas every four months to update information on migration (both internal and external), births, deaths, vaccinations and pregnancies. The NUHDSS has received ethical clearance from the Kenya Medical Research Institute's Ethical Review Board. Fieldworkers are trained on research ethics and obtain consent from targeted respondents before carrying out any interviews. Respondents who do not consent are not interviewed in this study.[3] Fieldworkers carry out rigorous data quality checks, including spot checks on five percent of the households, which are randomly selected.

3 Since 2005 when the NUHDSS started recording household level response rates, the refusal rates were 1.5 percent in 2005 and 2.0 percent in 2006. The common reasons given for refusal are lack of time (being too busy) or being tired of responding to the questions over a long time without seeing tangible benefits of the research to them personally or to the communities.

Table 8.1 Basic social and demographic indicators from NUHDSS, 2003–2006

Measures	2003			2004			2005			2006		
	Both Areas	**Koch**	**Viwa**	**Both Areas**	**Koch**	**Viwa**	**Both Areas**	**Koch**	**Viwa**	**Both Areas**	**Koch**	**Viwa**
Population as of 31 December	58,232	26,190	32,042	55,774	27,471	28,303	57,909	27,917	29,992	59,297	27,951	31,346
Total number of households on 31 December	23,380	9,029	14,351	22,423	9,500	12,923	23,087	9,671	13,416	23,424	9,587	13,837
Number of Births	1,753	802	951	1,889	993	896	1,791	913	878	1,941	932	1,009
Number of Deaths (all ages)	522	331	191	455	284	171	448	280	168	447	267	180
Sex Ratio (all ages)	133.3	111.5	155.6	129.7	110.6	152.4	129.0	111.0	149.2	126.6	109.7	144.7
Dependency Ratio (%)	44.4	59.2	33.8	46.8	60.3	35.6	47.3	60.7	36.6	48.8	61.6	38.7
Crude Birth Rate ('000)	32.4	32.3	32.4	35.3	37.3	33.4	32.2	33.7	30.7	34.5	34.3	34.6
Total Fertility Rate	3.1	3.4	2.9	3.5	3.9	3.1	3.2	3.5	2.9	3.2	3.5	3.1
Crude Death Rate ('000)	9.6	13.3	6.5	8.5	10.7	6.4	8.1	10.3	5.9	7.9	9.8	6.2
Infant Mortality Rate (per 1000 live births)	95.5	125.5	71.3	91.7	105.4	77.2	93.5	104.9	81.7	88.4	98.4	79.4
% In-migrants	22.0	16.9	26.1	26.5	20.2	32.6	22.2	17.4	26.7	19.6	16.9	21.9
% Out-migrants	17.5	14.3	20.1	35.5	20.5	50.1	20.8	18.1	23.3	18.9	18.2	19.5
Female headed household (% of total)	17.4	22.1	14.4	17.8	22.3	15.1	18.7	23.3	15.5	19.7	23.9	15.8
Average number of persons per household	2.4	2.7	2.1	2.4	2.8	2.2	2.5	2.9	2.2	2.6	2.9	2.3

Note: Koch=Korogocho; Viwa=Viwandani.

Both slum settlements are located about 5–10 km from the city centre and 3 km from each other. Each of the two settlements consists of seven villages. Although Nairobi is only 145 km (1.5 degrees) south of the equator, it has a moderate tropical climate, because of its high altitude of about 1,700 m above sea level. Table 8.1 presents the key characteristics of the study areas. At the end of December 2006, the population under surveillance was 56,946 in both locations with 27,160 living in Korogocho and 29,786 in Viwandani. There were a total of 21,371 households in the two locations, giving an average household size of 2.6 people. As noted above, one of the key features of slum settlements is congestion – typical households live in one-room houses where they sleep, cook and sometimes even bath. The study population is highly mobile with in- and out-migration rates of 17.6 percent and 15.5 percent, respectively in 2006. The two sites differ markedly in migration and a number of other indicators. Viwandani, which is located near the industrial area, involves more migration and has more people in salaried employment than Korogocho (Zulu et al. 2006). Korogocho's population is more settled, with stronger community cohesion than Viwandani. The average duration of stay in Korogocho is 14.1 years compared to 7.6 years in Viwandani and while 30 percent of the residents of Korogocho were born in the community or other parts of Nairobi, only nine percent fall in this category in Viwandani (Zulu et al. 2009).

Study Design and Analysis

Our population of interest is children who were born in the slum settlements between 1 January 2003 and 31 December 2006. The NUHDSS recorded a total of 7,374 births for the period covered by the study. However, after taking out cases with missing key variables and inconsistent data, we ended up with 6,998 births (95 percent of the total births recorded). The outcome of interest was the death of the children born in the slums during the study period, which means that children that were born in the slums before 1 January 2003 and those who were born outside the study communities but migrated to the communities during the observation period were not included in this analysis. The number of person years lived by the 6,998 children for whom we have complete data was 11,518 and 394 of them died by 31 December 2006. Thus, we are following up children born in the slums for a maximum of four years.

We used a Cox proportional hazards model to assess the effect of migration status of the mothers on childhood survival rates, after controlling for various demographic and socioeconomic factors. A child is considered to have been born to a recent migrant if he/she was born to a mother who migrated to the study community since January 2003. Children born to mothers who were in the study community before January 2003 are taken as children born to 'long-term residents'.[4]

4 It should be noted that most mothers who were in the study area when APHRC started running the NUHDSS are also migrants – 25 percent and 5 percent of the residents

So we are comparing death rates for children born in the slum settlements of which one group is born to recent migrants and the other group is born to those who were in the study areas before 2003; the so called 'long-term residents'. This allows us to measure the effect of parental migration that is not confounded by differences in exposure to other environments for the two groups since the children are born in the same slum context.

For each child, the observation time started at birth between January 2003 and December 2006 and ended either at the occurrence of the event of interest (death) or dates of censoring due to refusal, loss to follow-up, emigration or end of the follow-up when the observation time was truncated for the children who were still alive (31 December 2006). We allowed gaps in the observation time, meaning that children could out-migrate and come back and used a number of demographic and socioeconomic factors known to affect child survival as control variables. We have three types of control variables:

1. Those directly related to the child (age, sex, ethnicity and place of residence),
2. Those relating to the mother (age, survival status and level of education),
3. Those relating to the household economic status for the household where the child lives (source of water; whether they have own or shared/no toilet; roof type; floor type; ownership of dwelling unit, phone, radio and television).

One of the most critical issues that one should take note of and possible control for in longitudinal analyses is attrition, especially in cases where those leaving the study area are predisposed to different risks of dying compared to the general population. Attrition is particularly important in the NUHDSS setting because of the non-permanence of housing structures, unreliability of livelihood opportunities and the consequent high levels of population mobility both within and outside the location. The two main sources of attrition in the NUHDSS are out-migration and loss to follow-up. Analysis of data of people who migrated to the slum settlements between 2003 and 2007 shows that the median duration of stay for the new migrants was 22 months for males and 26 months for females in Korogocho and 18 months for females and 22 months for males in Viwandani (Zulu et al. 2009).

The NUHDSS field team sometimes fails to observe some individuals and households because it is hard to find an eligible respondent at home. The most difficult cases of loss to follow-up are named 'hanging cases', which are cases where the fieldworker confirms that the respondent has left the housing unit where he/she was living during the last visit and is informed that the person has moved to another location within the same slum. However, the fieldworkers fail to trace the person in the new location for several rounds. Because the effect of hanging cases

aged 12 years and above were born in Korogocho and Viwandani settlements, respectively (Zulu et al. 2009). The analysis would have been most intuitive if we had controlled for duration of stay in the slum location, but this information was not available.

is the same as out-migration (in that no events or updates can be done regarding the person), out-migration and exits are combined in one category to form the overall attrition. Out-migration is by far the most dominant source of attrition from the NUHDSS population. The NUHDSS data show that out of the 60,207 people (total population) who left the surveillance population between 2003 and 2007, out-migration accounted for 92 percent, while hanging cases and deaths accounted for five percent and three percent, respectively. Note, however, that the analysis on attrition uses the child as the unit of analysis as opposed to the mother.

All event history analyses make the explicit assumption of independence between censoring and event. When censoring is not independent from the event of interest (for example, migration in relation to death) then the results suffer from potential bias. In this analysis, we control for non-independent censoring and the consequent selection bias, that is, when the same determinants may cause attrition and mortality. We follow the rationale of two-stage regression models controlling for endogeneity by first modelling the attrition risk using available independent variables, including an instrumental variable (IV), for example, a variable that affects attrition but not mortality. Here we use the notice of demolition of household structures under the Kenya Power and Lighting Company (KPLC) electric lines that was given in 2004.[5] We then derive an individual, time-dependent propensity for attrition, using the cumulative hazards function. This function is preferred to the survival function because attrition is a repeatable event. Finally, we insert this propensity for attrition in the mortality model, as a time-varying variable, since it varies according to the time spent in the DSA and according to changes of individual characteristics during observation. The squared propensity term is also introduced in the model to account for non-linear effect of attrition.

Results

Table 8.2 presents the results from the analysis with the following three models:

1. The attrition model: note the use of the demolition notice as an instrument (determinant of attrition but not of mortality).
2. The base mortality model with covariates but no control for attrition.

---

5 For health reasons, KPLC issues a notice to all residents of Nairobi city whose houses were located below high voltage electricity lines to demolish their dwelling units within three months or face force eviction/demolition. This lead to mass demolition of houses in the two slums and while some of the residents relocated to other dwelling units within the two slums, a lot of the people from the affected houses moved out of the slums. That is why 2004 has a markedly high number of out-migrants than the other years (Table 8.1). While the notice of demolition clearly led to a higher level of attrition, there is no reason to believe that the departure from the study population due to this fact affected survival probabilities of the children beyond the normal effect of out-migration.

Table 8.2 Attrition and mortality models for slum-born children under four (2003–2006)

Variables	% of person years or mean value	Hazard Ratios (standard errors)		
		Attrition	Model A	Model B
Predictor				
Non-recent migrant mother [Ref.]	63.2%			
Recent migrant mother	36.8%	1.51 (0.057)***	1.35 (0.175)***	1.83 (0.231)***
Covariates				
Sex [Female=Ref.]	48.9%			
Male	51.1%	1.02 (0.034)	1.18 (0.120)	1.19 (0.122)*
Slum of residence [Korogocho=Ref.]	52.2%			
Viwandani	47.8%	1.08 (0.047)*	0.91 (0.119)	0.96 (0.126)
Mother's Ethnic group [Kikuyu=Ref.]	28.8%			
Kamba	19.6%	1.23 (0.063)***	0.76 (0.134)	0.89 (0.157)
Kisii	3.9%	1.10 (0.099)	0.67 (0.237)	0.73 (0.257)
Luhya	15.2%	1.22 (0.066)***	1.29 (0.204)	1.50 (0.242)**
Luo	20.7%	1.15 (0.059)***	1.70 (0.240)***	1.88 (0.269)***
Embu/Meru	2.0%	1.11 (0.139)	1.10 (0.431)	1.20 (0.469)
Other ethnic groups	9.8%	1.07 (0.085)	0.62 (0.174)*	0.65 (0.182)
Mother's Education [Primary=Ref.]	68.2%			
Non educated	7.2%	0.82 (0.075)**	1.52 (0.387)	1.28 (0.336)
Secondary and higher	23.5%	0.96 (0.041)	1.26 (0.156)*	1.22 (0.150)
Mother's age	25.08	1.01 (0.024)	0.98 (0.060)	0.98 (0.060)
Mother's age squared	-	1.00 (0.000)	1.00 (0.001)	1.00 (0.001)
Mother alive [Yes=Ref.]	99.4%			
No	0.6%	1.15 (0.231)	7.04 (1.525)***	7.54 (1.622)***
Presence of water tap [No=Ref.]	9.8%			
Tap water	90.2%	0.93 (0.053)	1.05 (0.201)	0.98 (0.189)
Household owns toilet [No=Ref.]	97.4%			
Owns a toilet	2.6%	1.01 (0.145)	0.56 (0.243)	0.56 (0.243)

Household has electricity [No=Ref.]	80.9%			
Has electricity	19.1%	0.90 (0.049)*	1.12 (0.171)	1.03 (0.160)
Owns a house [No=Ref.]	89.8%			
Owns a house	10.2%	0.67 (0.053)***	0.93 (0.182)	0.68 (0.148)*
Owns a phone [No=Ref.]	81.6%			
Owns a phone	18.4%	0.86 (0.048)***	0.95 (0.158)	0.85 (0.142)
Owns a radio [No=Ref.]	35.6%			
Owns a radio	64.4%	0.88 (0.031)***	0.77 (0.084)**	0.69 (0.076)***
Owns a television [No=Ref.]	83.1%			
Owns a TV	16.9%	0.79 (0.051)***	0.86 (0.153)	0.73 (0.128)*
Trimester	13.6	0.96 (0.004)***	0.94 (0.010)***	0.92 (0.011)***
Demolition notice [No notice=Ref.]	98.4%			
Notice	1.5%	2.97 (0.234)***	-	-
Attrition		-	-	
Attrition propensity	0.46			0.001 (0.001)***
Attrition propensity squared	-	-	-	20.18 (10.634)***
Number of person years	11518.38			
Number of children (events)		6998 (3639)	6998 (394)	6998 (394)

Note: Model A: Mortality model without controlling for attrition, Model B: Mortality model controlling for attrition. Robust standard errors computed for attrition model (repeatable events). Significance level: *=10%; **=5%; ***=1%. All models controlled for roof and floor type of the dwelling unit.

3. Mortality model controlling for attrition (using cumulative hazards computed from attrition model) and other covariates.

Attrition Model

The results show that being a recent migrant has a very strong and significant effect on the likelihood of leaving the study population during the study period (attrition). Children born to recent migrant mothers have a higher chance of attrition than children born to 'long-term residents' (HR=1.5, P<0.000). If the mother migrated in the slums for economic reasons and did not find a job or the child was not planned, then the woman is likely to go back home or send the child 'home'. In addition, unplanned children may be subjected to more health hazards than wanted ones, suggesting that out-migration may be health-related. The socio-demographic characteristics of the mother have mild effects on attrition while the household economic conditions have significant effects. In general, those with higher economic status (as measured by household possessions and amenities) are less likely to leave, suggesting that departure from the slums reflects failure to find livelihood opportunities and afford houses with relatively good amenities. Children living in households that own the dwelling units, phones, radios or television are significantly less likely to move out of the study area or be lost to follow-up. There are also significant differences in the likelihood of attrition across ethnic groups, with higher chances of attrition among the Kamba, Luhya and Luo ethnic groups compared to the Kikuyu. The Kikuyus' original home area (Central Province) is close to Nairobi and the lower levels of attrition in this group may be suggestive of the fact that many of the Kikuyus living in the slums are actually long stayers and more likely to be doing well economically. The notice of demolition effect is strong and significant (HR=3.0, $P<0.000$) on attrition but has no effect on mortality (results not shown). This confirms that notice of demolition is a good instrumental variable for our two-stage model.

Base Mortality Model (Model A in Table 8.2)

Children born to recent migrants have a significantly higher risk of dying than those born to 'long-term residents' and non migrants (HR=1.4, $P<0.01$). Children born to Luo mothers have a significantly higher risk of dying than children born to Kikuyu mothers. The mother's death is another variable with a strong and significant effect on child mortality (HR=7.0, $P<0.000$). Out of all the household economic measures, only ownership of the radio has a significant effect on mortality (deters child mortality). Notable variables that do not show a significant effect on child mortality in this model include shelter's slum location, mother's age, ownership of dwelling, ownership of television and household access to electricity, water supply and sanitation.

Mortality Model Controlling for Attrition (Model B in Table 8.2)

After controlling for the attrition effect, the results show that the attrition effect is very strong and convex (the squared term is significant), meaning that an increase of one unit of attrition propensity has stronger effect when this propensity is low (for example, increase from zero to one) than when it is already high (for example, increase from four to five). The results confirm that attrition is endogenous to mortality of the children: censoring is not independent from the death event and therefore is a major source of bias in estimating mortality. Mortality among children born to recent migrants is considerably higher than that of children born to 'long-term residents' after controlling for attrition (HR=1.8, P<0.01) versus (HR=1.4, P<0.01). The difference is somewhat underestimated without controlling for attrition since new migrants are associated with higher mortality and recent migrants are also more likely to out-migrate (see Zulu et al. 2009 for analysis of out-migration). This suggests that mortality rates in the slums would be higher if those who left had stayed in the population.

The effect of ethnicity remains strong (both the Luos and Luhyas have significantly higher mortality risks than Kikuyus). The effect of mother's death remains very strong (HR=7.5, P<0.000) after controlling for attrition. Most measures of socioeconomic status had no significant effect on child mortality after controlling for attrition except for ownership of a radio, which significantly reduces the risk of mortality (HR=0.6, P<0.000) and ownership of a television and a house, which marginally decrease the risk of mortality for children.

Discussion

Many studies have assessed the risk of infant and childhood mortality and associated risk factors (Ouoba 1998, Gyimah 2002, Mboup 2001). However, these studies are mostly based on cross-sectional data, known to be less appropriate for the analysis of most vital demographic events. The few studies that use longitudinal techniques are based on rural settings (for example, Becher et al. 2004) and the risk factors assessed rarely include migration.

The key studies that have specifically examined the effect of migration on child survival (for example Brockerhoff 1990, 1995, Stephenson et al. 2003, Bocquier et al. Forthcoming) have also been based on cross-sectional data, with limited migration data. In the present study we used rich longitudinal data collected from poor urban slum settlements in Nairobi City, Kenya, to assess the effect of the mother's migration status on childhood mortality. The main focus was to compare survival of children who are born in the same slum environment, according to their mother's migration status. This study goes a long way in enhancing understanding of the effect of migration on health outcomes and contributes to the understanding of the underlying factors for the relatively high disease burden exhibited by slum residents (APHRC 2002a, Ndugwa and Zulu 2008, Kyobutungi et al. 2008).

Our results show that children from recent migrant mothers are 1.8 times more at risk of dying than those from long-term resident mothers, after controlling for attrition and other factors. Before controlling for attrition, the effect of recent migration of the mother was smaller (1.4 times), demonstrating the importance of controlling for out-migration patterns in understanding child mortality. The result shows that out-migration is somewhat selective of people who are more prone to poorer health outcomes and that the mortality rates in slum settlements would actually be higher than the observed rates if many of the people who are leaving the locations stayed there longer.

The effect of migration status is underestimated without controlling for attrition since new migrants are associated with higher mortality and recent migrants are also more likely to out-migrate (see Zulu et al. 2009). Furthermore, attrition is lower for households with better economic indicators as measured by ownership of household assets (including ownership of dwelling units) and presence of amenities such as water and toilets. Another study from the same data indicates that new migrants are less likely to be employed and make less money (Zulu et al. 2006). The fact that the effect of migration status on child mortality remains strong and significant after controlling for attrition and household economic conditions is quite profound, especially when one considers that all the children considered in this study were born and observed in the same general (slum) environment. This calls for the need to understand why being new in the slum setting, other factors being equal, exposes children to higher risks of dying. The observed outcome may partly reflect lack of appropriate measurements of economic status and access to health care in the present study. However, evidence from other settings point to the adaptation problems that new migrants often face when they come to new settings. The new migrant mothers may not yet be integrated into the new environment and may not know how and where to seek health care services for their children. Kiros and White (2004) reported in Ethiopia low vaccination coverage (vaccination was provided free of charge) among children from migrant mothers compared to those from non-migrants and explained these differences by the integration level of the mother in the host community.

We also identified other major risk factors associated with childhood mortality in the slums. The significant effect of ethnicity on mortality suggests that there might be genetic and cultural factors that affect child mortality (Blacker 1991). Children born to Luo and Luhya mothers have significantly higher risks of dying than children born to Kikuyu mothers. We should also note that Kikuyus are more likely to be long-term migrants compared to Luos, Luhyas and other ethnic groups as shown in the attrition model. This could partly explain the mortality difference. This finding supports findings from other surveys like the DHS that show that Luos and Luhyas exhibit poorer health outcomes, including child mortality, than other ethnic groups in Kenya. These differentials call for detailed studies to understand the factors driving these huge ethnic differences in child mortality, especially that in our study, we are looking at children born and growing in the same slum context. Ethnic group differences in mortality risk have also been reported in other settings (Blacker 1991, Becher et al. 2004).

Most socioeconomic factors, with the exception of ownership of the radio, were not significantly associated with mortality risk. It is not clear why ownership of a radio stands out among other proxy measures of household wellbeing, especially that the analysis also included indicators of factors that may be typically associated with child mortality such as sanitation and water supply. However, possible explanation of the strong effect of radio ownership on mortality could be the access to information, especially health information.

Indeed, households with radio could listen to health promotion programmes that can change their health seeking behaviour positively. Unlike television, people can listen to the radio at anytime and anywhere including place of work.

Another major risk factor for childhood mortality is the death of the mother, which increased the risk of mortality. Similar findings have been seen in other studies (Koblinsky et al. 1994, Becher et al. 2004). Children who lose their mothers are likely to be exposed to several factors that can increase their mortality risk, including: reduction of care, cessation of breastfeeding, improper bottle feeding and HIV transmission from their mothers. Due to the HIV/AIDS epidemic an increasing number of children are experiencing the death of their mother and so the impact on childhood mortality will be substantial.

Limitations

Our data generated from the HDSS were appropriate to explore and assess risk factors associated with childhood mortality. A major limitation is that most of the variables included in our model were not time-dependent, except for age of the mother, trimester and notice of demolition. Trimester is meant to capture the trend over the period 2003–2006 and notice of demolition is used as an instrumental variable. For instance, we only used the household economic status data that were collected close to the date of birth for the child and these could have changed for the children observed for a long time. The NUHDSS started collecting data on household amenities and assets annually from 2006. This will allow treatment of these proxy measures of household economic wellbeing as time-varying variables in future analyses.

Conclusion

Childhood mortality in the Nairobi informal settlements remains very high, especially among new migrants. While emerging evidence highlight the need to pay attention to the plight of slum dwellers in African cities, this study demonstrates the need to look at inequities in health outcomes even within the so called 'marginalized groups'. Given the high degree of rural urban migration, which is bound to increase in the foreseeable future for most African countries if the large socioeconomic inequities between rural and urban areas are not addressed, it is important that policy makers understand and address factors predisposing children of recent migrants to such high risks of mortality. There is, therefore, need to investigate factors that predispose children born to recent migrants to higher mortality risks than those born in the same geographical environment but to 'long-term residents'. If poor utilization of health services by new migrants is the problem, as suggested in other studies, all efforts should be made to find cost-effective mechanisms for increasing access to health care for the recent migrants.

Acknowledgements

We would like to acknowledge the contribution of APHRC's dedicated HDSS field workers, team leaders, field supervisors and the Field Coordinator for their efforts in collecting the HDSS data in the field. We are also grateful to the data entry and management team for processing the data and making it available for the analysis. Especially, we would like to thank Zewdu Woubalem, Jacques Emina and Patricia Elungata for the remarkable preliminary data cleaning and processing work. We are also profoundly grateful to the residents of Korogocho and Viwandani slum settlements for taking their valuable time to respond to our questions when our field workers visit their households every four months. Last but not least, we acknowledge the generous support to the NUHDSS by funding from the Wellcome Trust, the Rockefeller Foundation and the Hewlett Foundation.

References

Agwanda, A.O., Bocquier, P., Khasakhala, A., Nyandega, I. and Owuor, S. (Forthcoming). *Biography of Three Generations of Nairobi Residents - Thirty Years of Social Urban History*. Dakar: CODESRIA.

APHRC 2002a. *Population and Health Dynamics in Nairobi Informal Settlements*. Nairobi Kenya: African Population and Health Research Centre.

Becher, H., Müller, O., Jahn, A., Gbangou, A., Kynast-Wolf, G. and Kouyaté, B. 2004. Risk factors of infant and child mortality in rural Burkina Faso, *Bulletin of the World Health Organization*, 82(4), 265–73.

Bocquier, P., Nyovani, M. and Zulu, E. (Forthcoming). The impact of urbanization and migration on child mortality in sub-Saharan Africa. *Demography*.

Blacker, J.G.C. 1991. Infant and child mortality: development, environment and custom, in *Disease and Mortality in Sub-Saharan Africa*, edited by R.G. Feachem and D.T. Jameson. Oxford: Oxford University Press.

Bogin, B. 1988. Rural-to-urban migration, in *Biological Aspects of Human Migration*, edited by C.G.N. Mascie-Taylor and G.W. Lasker. Cambridge: Cambridge University Press.

Brockerhoff, M. 1995. Child survival in big cities: the disadvantages of migrants. *Social Science and Medicine*, 40(10), 1371–83.

Brockerhoff, M. 1990. Rural-to-urban migration and child survival in Senegal. *Demography*, 27(4), 601–16.

Dodoo, F.N., Zulu, E.M. and Ezeh, A.C. 2007. Urban-rural differences in the socio-economic deprivation-sexual behaviour link in Kenya. *Social Science and Medicine*, 64(5), 1019–31.

Gould, W.T.S. 1998. African mortality and the new 'urban penalty'. *Health Place*, 4(2), 171–81.

Gyimah, S.O. 2002. *Ethnicity and Infant Mortality in Sub-Saharan Africa: The Case of Ghana.* Discussion Paper No. 02-10. London, Canada: University of Western Ontario: Population Studies Centre. Available from: http://www.ssc.uwo.ca/sociology/popstudies/dp/dp02-10.pdf (Last accessed 22 May 2009).

Kiros, G.E. and White, M.J. 2004. Migration, community context and child immunization in Ethiopia. *Social Science and Medicine*, 59(12), 2603–16.

Koblinsky, M.A., Tinker, A. and Daly, P. 1994. Programming for safe motherhood: a guide to action. *Health Policy and Planning*, 9, 252–66.

Kyobutungi, C., Ziraba, A., Ezeh, A.C. and Yé, Y. 2008. The burden of disease profile of residents of Nairobi's slums: results from a demographic surveillance system. *Population Health Metrics,* 6, 1. Published online at: http://www.pophealthmetrics.com/content/6/1/1; doi:10.1186/1478-7954-6-1. (Last accessed 22 May 2009).

Magadi, M.A., Zulu, E.M. and Brockerhoff, M. 2003. The inequality of maternal health care in urban sub-Saharan Africa in the 1990s. *Population Studies*, 57(3), 347–66.

Mboup, G. 2001. Mortality of children under five years of age (Mortalité des enfants de moins de cinc ans), in *Republique du Benin Enquete Demographique et de Sante*. Benin. Calverton (MD): Demographic Health Surveys, 115–24. Available from: URL: http://www.measuredhs.com/start.cfm (Last accessed 22 May 2009).

Mugisha, F. and Zulu, E.M. 2004. The influence of alcohol, drugs and substance abuse on sexual relationships and perception of risk to HIV infection among adolescents in the informal settlements of Nairobi. *Journal of Youth Studies*, 7(3), 279–93.

Ndugwa, R.P. and Zulu, E.M. 2008. Child morbidity and care-seeking in Nairobi slum settlements: the role of environmental and socio-economic factors. *Journal of Child Health Care*, 12(4), 314–28.

Ouoba, P. 1998. Mortality of children under five years of age (Mortalité des enfants de moins de cinq ans), in, *Enquete Demographique et de Sante, Burkina Faso*. Calverton (MD): Demographic Health Surveys, 135–44. Available in French from: URL: http://www.measuredhs.com/start.cfm (Last accessed 22 May 2009).

Prothero, R.M. 1977. Disease and mobility: a neglected factor in epidemiology. *International Journal of Epidemiology*, 6(3), 259–67.

Stephenson, R., Matthews, Z. and McDonald, J.W. 2003. The impact of rural-urban migration on under-two mortality in India. *Journal of Biosocial Science*, 35(1), 15–31.

UN Habitat. 2003. *The Challenge of Slums – Global Report on Human Settlements 2003*. London: Earthscan.

Zulu, E.M., Donatien, B., Mudege, N., Muindi, K. and Batten, L. 2009. *Characteristics of Recent In-Migrants in the Nairobi Urban Health Demographic*

Surveillance System. Paper presented at the Population Association of America Annual Meeting: Session 170, The Structuring of Internal Migration, Detroit, Michigan, 30 April–2 May 2009.

Zulu, E.M., Dodoo, F.N. and Ezeh, A.C. 2002. Sexual risk-taking in the slums of Nairobi, Kenya, 1993–1998. *Population Studies*, 56, 311–23.

Zulu, E.M., Konseiga, A., Darteh, E. and Mberu, B. 2006. *Migration and the Urbanization of Poverty in Sub-Saharan Africa: The Case of Nairobi City, Kenya.* Paper presented at the Population Association of America Annual Meeting: Migration in Developing Countries, Los Angeles, 30 March–1 April 2006.

Chapter 9

Child Migration and Mortality in Rural Nyanza Province: Evidence from the Kisumu Health and Demographic Surveillance System (KHDSS) in Western Kenya

Kubaje Adazu, Daniel Feiken, Peter Ofware, Bernard Onyango, David Obor, Rose Kiriinya, Laurence Slutsker, John Vulule and Kayla Laserson

Introduction

Problem Statement and Background

In Kenya, early childhood mortality is one of the critical challenges that public health practitioners and development agencies have been grappling with for the past one and a half decades. In spite of the introduction of Primary Health Care (PHC), the Expanded Program on Immunization (EPI) and the Integrated Management of Childhood Illness (IMCI), overall 77 infants out of every 1,000 live births die before the age of one year and of those who survive the first year of life another 41 out of every 1,000 die before the age of five (CBS Kenya 2004). Grim as the childhood mortality statistics at the national level look, the level of mortality in some of the provinces is even more troubling. For example in Nyanza Province, infant and under-five mortality rates increased from 94 and 149 deaths per 1,000 live births in the late 1980s to 133 and 205 deaths per 1,000 live births, respectively, a decade later (CBS Kenya 1990, 1999, 2004). In addition to the regional differences, there are also significant rural-urban differentials in child mortality rates. In general, children residing in urban areas have a better chance of surviving the first five years of life compared to rural children.

Situated in Asembo and Gem in Nyanza Province, one of the provinces with the worst child survival outcomes in Kenya, the demographic surveillance area (DSA) of the Kisumu Health and Demographic Surveillance System (KHDSS) has high rates of both in-migration and out-migration for children from and to neighbouring and distant villages, towns and cities. For instance, 15 percent of the under-five population in the DSA in 2003 experienced at least one migration event. The DSA is predominantly rural and characterized by persistent high infant and child mortality (Spencer et al. 1987, McElroy 2001, Adazu et al. 2005). Most of the early childhood deaths are caused by infectious diseases, with malaria, pneumonia, anaemia and malnutrition being the major causes of death. In addition to the infectious disease burden, access to safe water is limited and diarrhoeal diseases are important contributors to morbidity and mortality amongst children.

The existence of urban-rural child mortality differentials in favour of urban areas in Kenya coupled with the morbidity burden that characterizes the DSA suggests that migrating from urban settings into the DSA could be detrimental to the health and survival of young children. Children migrating into the DSA from low mortality settings may be exposed to new disease pathogens that their immune systems have not yet developed any resistance against and without any immunity the chances of dying in the event of an attack are high. The reason for returning to the rural area may be linked to illness, including HIV of mother or child. The risk of death shortly after arrival at the destination is high in such cases, especially if the migrants are terminally ill (Clark et al. 2007). Dissolution of marriages due to divorce or widowhood, one of the reasons for migration in the DSA, sometimes leads to the sudden departure of mothers and their young children from the matrimonial home with the attended consequence of both mothers and children being deprived of the economic and social support needed to bring up healthy children.

It is however plausible that the children migrating into the DSA are positively selected; in other words, only the healthy surviving children of migrant parents are brought into the DSA. Perhaps, in-migrants into the DSA time their return to coincide with the least malaria transmission periods thereby minimizing the chances of contracting malaria and the associated risk of dying from it. Furthermore, migrant families are typically made up of highly motivated and innovative individuals who tend to be more proactive; seeking employment and better lives also means such parents are more likely to vaccinate their children and seek medical care when their children are sick. The probability of death in childhood is significantly reduced for children who are timely and fully immunized (Rutstein 2000, Claeson and Waldman 2000).

Whether child migration is really implicated in mortality in the DSA is open to speculation because no study has yet investigated the relationship between these two demographic processes in the DSA. The amount of child migration in this part of Kenya is substantial and knowledge of its impact on child health outcomes could inform child health programs and interventions. Our main objective in this study therefore is to evaluate the relationship between child migration and mortality by examining the level of mortality amongst migrant and non-migrant children in the DSA.

Migration and Child Mortality

Research on the impact of migration on child health outcomes is largely focused on the role of parents' migration, especially the migration status of mothers, as mothers are generally more involved in childcare (Brockerhoff 1990, 1994, 1995, Ssengonzi et al. 2002, Stephenson et al. 2003, Kiros and White 2004, Kevin and Thomas 2007). The findings of these studies suggest that the relationship between migration and child health is complex, depends on the context and varies over time and among different kinds of migrants. For instance, using the Demographic and Health Survey data from 17 countries, Brockerhoff (1995) reported higher levels of child mortality among internal migrants in Mali, Senegal and Togo but lower child mortality among internal migrants in Ghana and Uganda. In an earlier study in Senegal, Brockerhoff (1990) compared the child mortality rates of rural-urban migrants and rural non-migrants and noted that the children of the rural non-migrants had higher mortality. Ssengonzi and his colleagues used the 1995 Ugandan Demographic and Health Survey (UDHS) data to investigate the impact of migration on child survival but in this study they considered rural-rural, urban-urban and urban-rural migration streams in addition to rural-urban migration. The evidence from this study showed that overall, migration status was marginally associated with child survival. Migrant status had the strongest positive impact on children of urban-to-urban migrant mothers and was not significantly related to the survival of children of rural-urban migrant mothers, contrary to what was found by Brockerhoff. The data set used by Ssengonzi and his colleagues was more recent, suggesting a probable deterioration in health outcomes for the children of rural-urban migrant mothers or random variation in the samples selected for the two surveys.

Using the 1996 Census data for South Africa, Thomas (2007) went beyond comparing child mortality differentials of migrant and non-migrant groups and in addition investigated mortality differentials between different socioeconomic groups within the internal migrant, immigrant and native born non-migrant populations. The results of this study suggests that although the children of internal migrants had better chances of surviving than the children of non-migrants, the survival advantage varied across different socioeconomic groups; the advantage of internal migrants over non-migrants was greatest in the wealthiest quintiles and least in the poorest quintiles. In a similar study in Bangladesh, Islam and Azad (2008) also observed that within urban areas child survival outcomes were worst among children of poor rural-to-urban migrant mothers, especially among recent migrants.

Three mechanisms, namely disruptive effects of migration, adaptation by migrant children and their households to the new environment at the destination after the move and migrant selectivity have been used to explain migrant and non-migrant child survival differences (Brockerhoff 1994). The disruptive effects are postulated to operate through disruption of social networks and childcare, change in diet, ingestion of contaminated food or water during the migration process. In addition to the hardships associated with the movement, on arrival at the destination migrant children and their mothers may also face difficulties fitting into the new communities due to language and cultural differences which could hamper their access to and utilization of healthcare services. The parents of migrant children may also have difficulty finding jobs at the destination. Inability to find jobs immediately upon arrival may lead to the depletion of personal savings which in turn may lead to migrant parents not being able to adequately feed and cloth their children. Indeed some studies have shown that children migrating from rural areas to urban settings experience increased mortality risk shortly after relocating (Brockerhoff 1994).

The adaptation effects operate through duration of residence at the place of destination, the receptivity of the host population and institutions and to some extent the circumstances that have given rise to the migration. This hypothesis postulates that over time migrants change their behaviour resulting in a convergence of health outcomes of both migrants and non-migrants. Using the 1986 DHS data for Senegal, Brockerhoff however showed that while children of urban migrants had better chances of surviving than children of rural non-migrants, compared with the non-migrants at the urban destination, the children of the urban migrants had higher risk of mortality, suggesting some but not total convergence in the case of Senegal. In another case study in Bangladesh, Islam and Azad (2008) also noted some evidence of adaptation among rural-urban Bangladeshi migrants. The results of this study showed that whilst infant and child mortality rates for both recent and long-term rural-urban migrant were higher than the rates for non-migrant urban residents, the mortality rates for the recent migrants were much higher than the corresponding rates for the long-term migrants.

The selectivity effects influence survival through household and individuals traits that are both associated with the propensity to migrate and child survival. However, the extent to which migrant selectivity predisposes child migrants to higher or lower mortality risks depends on the context in which the migration is taking place. There are two broad categories of child migrants; those who migrate with their parents and those who migrate without the parents. Those migrating without the parents are usually fostered children. Child fostering by its nature could have consequences for child survival, consequences that may be positive or negative depending on the social circumstances that have given rise to fostering of the child, the treatment the fostered child receives outside the parental home and the age of the child at the time of fostering. Street children or older children leaving home in search of work are other groups of children who migrate without their parents. Child labour or homeless children are however not the focus of this study hence the survival chances of these children in relation to their migration experience will not be discussed here.

The migrations of the children who are moving with their parents are occasioned by the parents' desire to migrate. Thus the survival chances of these children would largely depend on the underlying selection factors that are associated with the parents' migrations. Many case studies of rural-urban migration in Africa and other developing countries have shown that rural-urban migrants are generally young and most often better educated, compared with rural non-migrants (Marcos and Dacunha 1998, Zhu 1998, Guest 1998). This type of selection usually referred to as 'positive selection', generally occurs in situations where migrants voluntarily leave home in response to real or perceived opportunities ('pull factors') at the places of the destinations. High socioeconomic status and education, particularly high maternal education, enhance child survival (Cleland and Ginneken 1988, Bicego and Boerma 1993). Therefore the children of these migrants are likely to experience lower risk of death compared with the children of the poor and uneducated non-migrants in the countryside. Other studies have also documented negative selection, the situation where the migrants are predominantly of low socioeconomic status with little or no education; attributes highly associated with higher childhood mortality. This usually happens in situations where the migrants are leaving the places of origin in an attempt to escape from natural calamities such as drought or famine. For example, high infant and child mortality and morbidity rates were reported for the migrants who fled the Sahel draught in the 1970s (Caldwell 1975, Colvin 1981) and the famine in Ethiopia in the 1980s (Shears and Lusty 1987). The self selection among the children migrating with their parents could also arise from the characteristics of the children. For example, rural-urban migrant parents with frail children may relocate to the rural homes in order to reduce cost of care.

Besides the socio-demographic and behavioural factors, migration could also alter the survival chances of children through exposure to new disease pathogens and other environmental risk factors. The environment in which children are raised is known to have an impact on their health and survival (Pongou et al. 2006, Hoque et al. 1999). In addition, the environmental health risk factors vary significantly across geographic and climatic regions. For instance, data from the demographic health surveys have repeatedly shown significant infant and child mortality differentials between rural and urban areas. Several other cross-sectional studies have also documented an association between place of residence and childhood mortality (Sastry 1997a, Defo 1996). Within urban areas, mortality rates for children residing in informal or slum dwellings are much higher than the rates of children living in formal settlements (Madise and Diamond 1995, APHRC 2002). Thus the survival chances of child migrants may be enhanced or compromised as a result of the migration depending on the degree to which the disease burden at the places of origin and destination vary.

Migration could also influence child survival outcomes through access or lack of access to healthcare services, especially in the settings where healthcare facilities are concentrated in urban areas. For example, the persistent rural-urban child mortality differentials in Sub-Saharan Africa are often attributed to the availability of public health resources such as healthcare facilities and other social amenities which are beneficial to child survival in urban areas (Brockerhoff 1990, Sastry 1997b). On the basis of that explanation, rural-urban migration could enhance the survival chances of child migrants. However, migrating into urban shantytowns or informal settlements where healthcare services, housing, water and sanitation are inadequate could compromise the health and survival of children. Conversely, urban- rural migration could be detrimental to the health and survival of child migrants given the poor access to healthcare in many rural areas.

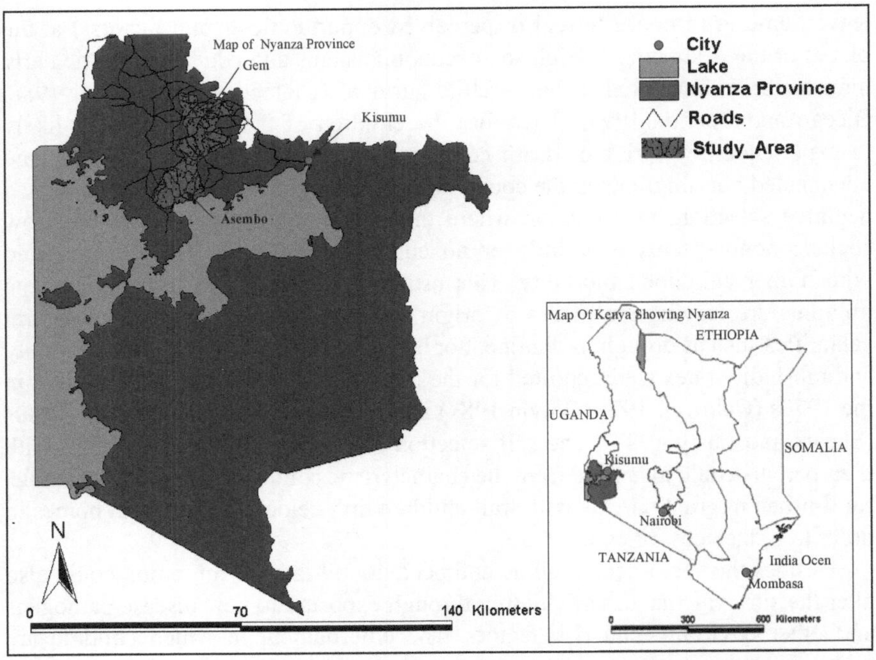

Figure 9.1 Map of the study area – DSA, Nyanza Province, Western Kenya

The Study Area

The Demographic Surveillance Area (DSA) is approximately 40 km away from the city of Kisumu in the Nyanza Province. It is located in Asembo in the Rarieda division of Bondo district and in Gem, which is situated in the Saiya district (Figure 9.1). The DSA covers a total of 217 villages (75 in Asembo and 142 in Gem), spread over a land area of about 500 km^2. In April 2007, the study area was expanded to include all the 168 villages in the Karemo division in Siaya district bringing the total number of villages to 365. The population is predominantly rural, culturally homogeneous (over 95 percent are members of the Luo tribe) and lives in dispersed settlements. Subsistence farming is the mainstay of the local economy. Rainfall is seasonal with the heaviest rains usually falling from March through May followed by a short dry period in August and a minor wet season between September and November. The crops cultivated for local consumption include maize, sorghum and cassava. Some households also raise poultry, goats, sheep and cattle and others engage in petty trading. Poverty is widespread and employment opportunities are limited, partly contributing to the out-migrating of young adults to the urban areas to seek employment.

The Kisumu Health Demographic Surveillance System (KHDSS)

The Kisumu Health and Demographic Surveillance System (KHDSS) was launched in September 2001 by the Centres for Disease Control and Prevention (CDC) and the Kenya Medical Research Institute (KEMRI) with the objective of providing timely demographic, morbidity and socioeconomic data. Details of the KHDSS design and methods have been published elsewhere (Adazu et al. 2005). The core of the KHDSS consists of house-to-house interviews conducted on a rolling basis through three rounds in each calendar year. The first round of data collection each year runs from January to April, the second from May to August and the last round from September to December. During each round of data collection, a team of trained community interviewers visit every household in the DSA to record morbidity episodes that occurred two weeks prior to the interview and demographic events (pregnancies, births, deaths and migrations) that occurred after the last visit. In addition to these surveys, household socioeconomic and educational status surveys are conducted annually to complement the morbidity and demographic data.

Only individuals who have resided in the DSA for at least four calendar months or children born to resident mothers are registered. At the time of registration, all residents and households are assigned unique identification numbers that makes it feasible to link individuals to their respective households and children to their parents, if the parents are registered in the KHDSS.

The KHDSS was reviewed and approved by the institutional review boards of both the Centres for Disease Control and Prevention in Atlanta (GA, USA) and the Kenya Medical Research Institute in Nairobi (Kenya). The heads of all compounds participating in the surveillance gave informed written consent for the participation of their families in the KHDSS.

Methods

Definition of a Migrant

In the context of the KHDSS, the boundaries of the study area constitute the boundary for defining migration. An individual is considered to have out-migrated if that individual was previously registered in the HDSS but moved to a destination outside of the study area and stayed there for at least four calendar months. An in-migrant is someone who has moved into the study area from a place outside of the study area and remained in the study area for at least four calendar months. This individual could be someone who was previously registered in the KHDSS and moved away or someone who is coming into the area for the first time.

The KHDSS also tracks changes in places of residence within the boundaries of the surveillance area and these types of movements are called transmigration (local mobility). A trans-migrant is an individual who moved from one compound to another within the DSA and stayed in the destination compound for at least four calendar months.

For each move, data are collected on the reason, date and the nature of the origin and destination.

Data and Methodology

For this analysis we selected all children born in 2004 and observed each child from the time he/she came into the surveillance area up to 31 December 2006 or until the day he/she exited from the area through death or out-migration. We linked each child to the parents, where possible, and the household in which the child is registered using the permanent identification numbers and extracted information on the characteristics of the parents and the households. We then used the date of entry into the KHDSS and the last date of observation (exit date for those who died or left and 31 December 2006 for those who were alive and present) to compute duration of exposure for each child. The duration window was divided into sub-spells of 120 days and the data transformed from one record per child to multiple records per child. For children who migrated out of the DSA and returned at a later date earlier than 31 December 2006, the periods spent outside of the surveillance area were excluded from the analysis.

We used multivariate event history models to investigate the relationship between child migration and mortality conditioned on surviving the first three months of life. The advantage of event history models over other conventional regression models is their capacity to handle censoring. Over a quarter of the children in the sample selected for the study out-migrated before the end of the observation period (31 December 2006). Thus their survival status as of 31 December 2006 was not known to us. Simply treating all these cases as lost-to-follow-up (LTFU) and excluding them from the analysis could severely bias the parameter estimates. Of those who did not migrate, 80.4 percent were alive and were also censored as at the end of the observation period. With this huge number of censored cases it is critically important to select a model that is capable of handling censoring.

We chose the time piece-wise constant exponential hazard model to examine the relative risk of death between migrant and non-migrant children in the DSA. This model takes into consideration duration and it does not require the hazard of death to be constant nor monotonic. With this model, the period of observation is divided into smaller intervals. The hazard is allowed to vary across intervals but within an interval the hazard is assumed to be constant. This model was particularly suitable for our data because in addition to the huge number of censored cases, the duration of residence in the DSA varied between migrants and non-migrants.

Variables

The dependent variable in the multivariate analyses was the hazard of death, which was modelled as a function of migration, the sex and age of the child, maternal age and education, household socioeconomic status, home-based water treatment and source of income. Migration was treated as a time-varying covariate and represented by two dummy variables, urban-in-migrant and rural-in-migrant. This allowed the switching of migration status from one category to another thereby making it feasible to capture the temporal sequence of the migration history of every child. For instance, a child that came into the DSA through birth and migrated to an urban destination for a while and came back, then out-migrated again to a rural destination and finally returned to the DSA, contributed person time to the non-migrant group, the urban-in-migrant group as well as the rural-in-migrant group. This child was classified as a non-migrant from the time of birth up to the time he/she out-migrated, then classified as urban-in-migrant from the time he/she came back from the urban area and rural-in-migrant from the time he/she came back from the rural area.

Sex is a categorical variable and was represented by a dummy variable which was coded as one if the child was a boy and zero otherwise. The age of the child (measured in months) was treated as a continuous time-varying covariate. The duration window was divided into sub-spells of 120 days with the age of the child computed as at the beginning of each sub-spell. The age of the mother was also treated as a categorical variable (teenage mother versus non-teenage mother) and represented by a dummy variable in the models.

Although level of educational attainment is quite low among the resident population of the DSA, only a small number of people have never attended school. As a result there were very few mothers in the sample that had no education at all. We therefore combined the mothers with no education with those who had only primary education into one category and mothers with post-primary school education into another category. In the multivariate models, children whose mothers had primary or no education were used as the reference category.

We represented household socioeconomic status by a dummy variable, namely thatch roof and/or mud house or otherwise. By Kenyan standards, corrugated zinc roofing sheets (called mabati in the local parlance) are quite expensive and those who can afford a mabati roof are generally considered to be of higher socioeconomic status while thatch roof is associated with lower socioeconomic status. Brick or cement-block houses are also associated with high socioeconomic status whilst mud houses are generally associated with low socioeconomic status. Also included in this variable was whether the house was a permanent or semi-permanent structure. We also included the household head's source of income: farming, salaried employee, small business or other.

We considered two aspects of drinking water: whether it was from a safe source (for example, boreholes, protected wells) or not and whether the water was treated with chlorine and/or boiling or left untreated.

Results

Descriptive Analyses

A total of 6,658 children born in the year 2004 were registered in KHDSS by 31 December 2006. Out of this number 1,688 (25.3 percent) were born elsewhere and later migrated into the KHDSS and the rest were born in the DSA. The age at which the migrant children came into the DSA varied from zero to 35 months with a mean of 12 months and standard deviation of nine months (Table 9.1). The minimum entry age for the children born in the DSA was greater than zero because there were four children who were registered several months after birth. One was registered six months after birth, one after 20 months, one after 23 months and the last one was registered after 25 months. The mothers of these children were frequently moving around within the DSA. As a result they did not meet the residency criteria to be registered in any compound until after several months. By the end of the observation period (31 December 2006), 27.5 percent of the total number of children had out-migrated underscoring the high mobility in the area and 14.2 percent had died (Table 9.2). Among the children who came in through migration, 742 (43.9 percent) out-migrated, 106 (6.3 percent) died and 840 (49.8 percent) were alive and still residing in the DSA. Of those born in the DSA, 1,088 (21.9 percent) out-migrated and 842 (16.9 percent) died. We further stratified the in-migrants by place of origin. Lower mortality was observed among the children who came in from urban origins (5.0 percent) compared with those who in-migrated from rural origins (8.1 percent), the difference in mortality between the two migrant groups was, however, not statistically significant.

The children were evenly distributed by sex across the migrant and non-migrant groups (Table 9.3). The majority of the children were living in houses that were constructed wholly or in part with earthen materials, reflecting the high level of poverty in the DSA. For instance, 55.5 percent of the children were living in mud houses roofed with grass-thatch and 25.3 percent were living in mud house that were roofed with mabati (zinc) and only 18.1 percent were reported to be living in permanent or semi-permanent houses. From the results in Table 9.2, mortality was slightly higher among children who were residing in mud houses or houses roofed with thatch.

The results showed a strong association between high maternal education and child survival. For instance, the proportion of deaths among the children whose mothers had at least secondary education was 7.8 percent, approximately half of that for the children whose mothers had no education (17.6 percent).

Teenage mothers contributed 25.4 percent of the total number of children. The date of birth of 165 mothers could not be determined because these women were not resident members of the DSA, thus it was not possible to determine the ages of these women. The proportion of children of teenage mothers lost to follow up was approximately 45 percent, twice that of children of non-teenage mothers (21.5 percent), underscoring the high prevalence of out-migration among female teenagers in the surveillance area. The deaths recorded in the observation period for children of teenage mothers and mothers aged 20 years and above were 213 and 735, accounting for 12.6 percent and 15.1 percent of the children in each of the respective age categories. In the undetermined category, ten deaths were reported which accounted for 6.1 percent of the children in this category.

Access to portable water is limited in the DSA; only 16.2 percent of the children selected for this study came from households that draw water from safe sources such as boreholes and protected wells. In spite of the unsafe sources from which the majority of the residents drew their water, a very small proportion of the children (13.5 percent) came from households that treated their water with water guard or other chlorine-based compounds before consumption. The proportion of deaths among the children residing in households that used untreated water was higher (15.2 percent) than that for children who were residing in households that used water treated with water guard or chlorine (11.9 percent).

Regular access to cash income was also a good predictor of mortality. The results showed that 8.2 percent of the children from households where the head was reported to be a salaried employee died compared with 14.3 percent for those from households where the head was reported as a farmer or a small businessman (15.8 percent).

Multivariate Analysis

In Table 9.3 we present the results of the multivariate event history models that we employed to access the relation between child migration and mortality net of duration of residence at the DSA and other socio-demographic risk factors. In these analyses we excluded all deaths that occurred in the first three months of life. The association between urban-rural migration and mortality observed in the descriptive analysis persisted after controlling for the duration of residence in the DSA (Table 9.3, Model I). The differences in the risk of death between urban-rural migrant and non-migrant children in this model were large and statistically significant. For instance, the hazard of death was reduced by 31 percent for children whose origins of migration were reported as urban. The difference in risk of death for rural in-migrants was not significantly different from that for non-migrants. We then controlled for the effects of being a teenage mother and the sex of the child by incorporating these variables in the analysis (Model II). The sex of the child was not a good predictor of mortality, nor was whether or not the mother was a teenage mother. Compared with the results in Model I, the direction and level of significance of the association between urban-rural migration and mortality did not change after controlling for the child's demographic risk factors. After controlling for the demographic characteristics of the child and mother, children who came into the DSA from rural origins had a higher risk of death compared with non-migrant children. In this model, rural migration was associated with a six percent increase in the risk of death, although the difference in risk between the non-migrant and rural-in-migrant children was not statistically significant.

The socioeconomic status of a child's household, the immediate environment in which the child is being raised and the availability of safe water in the household are important proximate determinants of child health outcomes. To test whether these factors modified the child migration–mortality relationship, we further incorporated the level of education of the mother, the type of house in which the child resided, home-based water treatment and the proxy variables for occupation of the household head in our analysis (Model III). There was a slight increase in the hazard ratios for the migration variables but the overall migration–mortality relationship did not change when we controlled for these other risk factors. The migrant children who came from urban origins had a lower risk of death whilst those who came from rural origins had higher risk of death (although the latter was not statistically significant). The results of Model III also show that maternal education has a significant positive effect on the survival of children. The risk of death for children of mothers with at least secondary education was 44 percent less than that for children whose mother had either primary education or no education at all.

Consumption of untreated water significantly increased the risk of death among children in the DSA. For instance, the risk of death for children from households that were reported to be using untreated water was 23 percent higher than that for children from households that treated their water. Regular access to cash income was found to be a good predictor of mortality. Being the child of a salaried employee versus a small business operator or farmer household head was associated with 35 percent reduction in the risk of death. Surprisingly, housing structure turned out not to be associated with child mortality.

Discussion and Conclusion

Most of the research on child migration has focused on older children who left home to work in settings that are potentially hazardous or exploitative or on the migrant behaviour of the parents. In this study we investigated the survival chances of very young migrant children in a rural population in western Kenya, using data from the Kisumu Health and Demographic Surveillance System (KHDSS), an ongoing longitudinal surveillance survey. The results showed that child migration was not a risk factor for mortality. Indeed, in-migrant children from urban areas had a lower risk of mortality compared with non-migrant children. The migrant survival advantage persisted after we controlled for the age of the child and household socioeconomic status. The relative low risk of death among the urban-migrant children might be partly due to migrant selectivity. It is plausible that most of the frail children of urban-rural migrants died in the urban areas leaving only the robust healthy survivors to migrate with their parents into the DSA.

Almost all migrant children came into the DSA in the company of their mothers. It may be that differential child raising practices between migrant and non-migrant mothers also contributed to the differences in the risk of death. Exposure to urban settings likely increases access to the mass media, which in turn increases access to information. Access to information on child-care likely enhances knowledge and a better understanding of the causes of early childhood mortality. Being exposed to the urban environment, urban-to-rural migrant mothers are more likely to have had access to information on better child-care. Urban-rural migrant mothers may therefore be more likely to adopt modern child raising practices, cleaner food preparation and storage and improved hygiene practices compared with rural non-migrant mothers. Consequently, urban-rural migrant mothers may be more likely to seek modern medical care for their children in the event of ill health. We were however unable to assess this in our data or tease out the mechanisms through which migration acted to influence mortality in this study due to the limited covariates in our data.

Numerous studies have repeatedly shown that children whose mothers are educated are less likely to suffer from malnutrition (Wolfe and Behrman 1982, Barrera 1990, Thomas et al. 1990, Desai and Alva 1998), more likely to be fully and timely immunized (Pebley et al. 1996), more likely to be given oral re-hydration therapy when they have diarrhoea (Coreil and Genece 1988, Galvao et al. 1994) and are less likely to die in infancy (Guilkey and Riphahn 1998, Blakely et al. 2003). The results of this study also attest to the importance of the mother's education to child survival. Maternal education stood out as a strong predictor of childhood mortality net of migration and the age of the child. Being the child of a mother with post-primary school education was associated with over 40 percent reduction in the hazard of early childhood mortality.

Although safe water interventions are being rolled out in the DSA, the lives of many children are still at risk due to lack of access to clean water. Four out of every five of the children in this study came from households that were reported to be drawing water from sources that were likely to be contaminated. More disturbing, the consumption of untreated water was associated with a significant increase in the risk of death. Controlling for migration status, age and other household socioeconomic factors like housing structure, the risk of death for children from households that were reported to be using untreated water was 30 percent higher than that for children from households that treated their water with chlorine-based compounds. The results of this study have also highlighted the importance of safe water and adequate sanitation for the survival of children and the urgent need to roll out more safe water interventions and educational campaigns on water treatment and storage. In settings like the DSA where pipe water is nonexistent, home-based water treatment and safe storage techniques have been shown to improve water quality and dramatically reduce the incidence and prevalence of diarrhoea and childhood deaths associated with diarrhoea (Crump et al. 2005). Thus, promoting home-based water treatment will no doubt help in improving the survival chances of children in the DSA.

A household's ability to provide for the health needs of its members and to mobilize resources in times of life-threatening emergencies depends on the financial situation of the head, especially in households where the head is the sole breadwinner. Of the proxy variables for the occupation of the head of the household in our analyses, salaried employee was associated with lower mortality, underscoring the importance of access to regular cash income in a predominantly subsistence economy as in the surveillance area. Our data did not allow us to explore the mechanism(s) through which the main source of income of the household head influences the survival chances of children in the DSA. However, evidence from studies done in similar settings in Africa and Asia have shown that utilization of preventive services such as immunization (Debpur et al. 2005) and curative services such as clinics and hospitals (Khe et al. 2005) are associated with households' economic circumstances. Easy access to both preventive and curative healthcare service and improved nutrition are therefore plausible explanations for the survival advantage of children from households whose heads were salaried employees.

This study has demonstrated the value of collecting longitudinal surveillance data and the value of such data for exploring and testing hypotheses. However, one major drawback of the surveillance data is that it has information on a very small number of covariates. Besides the limitation associated with the depth of information being collected, cost constraints have also limited its coverage to relatively small geographical areas hence rendering the information generated nationally unrepresentative. It is worthy considering expanding the number of covariates to include more socioeconomic and socio-demographic traits such as parity and marital status and population-based information on healthcare and health services utilization. The collection of a wider array of socioeconomic and health data in a surveillance setting like the KHDSS, combined with the intrinsic temporal richness of the longitudinal data, could significantly contribute to our understanding of the factors influencing child health outcomes. Very recently, the KHDSS started collecting data on marital status and more covariates are planned for inclusion in the KHDSS. The additional information could also be collected through special surveys so as not to overload the surveillance data collection machinery.

To revisit the motivation question of this study: do children who migrate from elsewhere into the DSA have elevated risk of early childhood mortality compared with non-migrant children? The empirical evidence from the data we used for this study does not support this, especially not for children who have moved into the surveillance area from urban settings. Further analyses on cause specific mortality and health seeking behaviour among migrant and non-migrant households are being done and the results of these analyses will help us better understand the relationship between migration and child mortality.

Acknowledgements

The authors gratefully acknowledge the contribution of the community interviewers, the KHDSS field supervision team, data clerks and managers and the administrative and transport staff of the KEMRI/CDC Program. We also acknowledge the support of the local administration and the chiefs and residents of the surveillance area. Finally we wish to thank Cheikh Mbacké, John Oucho and the Scientific Advisory Committee of the Migration and Urbanization Working Group for their insightful comments and the INDEPTH Secretariat for arranging for external reviewers to review this chapter. The KHDSS is a member of the INDEPTH network. This chapter has been approved by the Acting Director of KEMRI.

Table 9.1 Summary entry-age statistics, by entry type

Entry type	Entry Age (in months)				
	No. of Children	**Mean**	**Std**	**Min**	**Max**
Birth	4,970	0.01	0.59	0	25
Rural-in-migration	692	12.21	8.96	0	35
Urban-in-migration	996	12.06	9.12	0	35
Total	**6,658**				

Table 9.2 Distribution of the children by selected background characteristics

Variable	Total	Present	Out-migrated	Died
	Number	*Number (%)*	*Number (%)*	*Number (%)*
Entry Type				
Birth	4,970	3,040 (61.2)	1,088 (21.9)	842 (16.9)
Migrant				
Rural in-migrant	692	349 (50.0)	290 (41.9)	56 (8.1)
Urban in-migrant	996	494 (49.6)	452 (45.4)	50 (5.0)
Sex of Child				
Female	3,295	1,915(58.1)	915 (27.8)	465 (14.1)
Male	3,363	1,965 (58.4)	915 (27.2)	483 (14.4)
Housing Structure				
Mud	5,384	3,156 (58.62)	1,430 (26.6)	798 (14.8)
Permanent	555	337 (60.1)	158 (28.5)	60 (10.8)
Semi-permanent	653	353 (54.1)	222 (34.0)	78 (11.9)
Missing	66	34 (51.5)	20 (30.3)	12 (18.2)
Roofing Material				
Mabati (corrugated zinc sheets)	2,886	1.601 (55.5)	936 (32.4)	349 (12.1)
Thatch	3,706	2,245 (60.6)	874 (23.6)	587 (15.8)
Missing	66	34 (51.5)	20 (30.3)	12 (18.2)

Level of Education of Mother				
No Education	216	132 (61.1)	46 (21.3)	38 (17.6)
Primary	5,103	2,971 (58.2)	1,347(26.4)	785 (15.4)
Secondary	859	544 (63.3)	248(28.9)	67 (7.8)
Missing	480	233 (48.54)	189 (39.4)	58 (12.1)
Age of Mother				
Teenager	1,693	719 (42.5)	761 (45.0)	213 (12.6)
Non-teenager	4,965	3,161 (63.7)	1,069 (21.5)	735 (14.8)
Source of Drinking Water				
Safe Source	1,080	636 (58.9)	304 (28.1)	144 (13.3)
Unsafe source	5,578	3,244 (58.2)	1,526(27.4)	804 (14.4)
Home-Based Water Treatment				
Chlorine	896	552 (61.6)	216 (24.1)	104 (11.6)
Boiling/Filtration	2,272	1,294 (57.0)	667 (29.4)	311 (13.7)
Untreated	3,490	2,034(58.3)	947 (27.1)	533 (15.3)
Household Head Source Of Income				
Farming	3,894	2,193(56.3)	1,145 (29.4)	556 (14.3)
Salaried employee	328	215 (65.5)	86 (26.2)	27 (8.2)
Small business	951	580 (61.0)	221 (23.2)	150 (15.8)
Any other business	1,485	894 (60.1)	378 (25.4)	215 (14.5)
Total	**6,658**	**3,880 (58.3)**	**1,830 (27.5)**	**948 (14.2)**

Table 9.3 Distribution of relative hazard of death by migration status and other selected covariates (excluding deaths in the first three months of life)

	Model I		**Model II**		**Model III**	
Variable	**Haz. Ratio**	**CI (95%)**	**Haz. Ratio**	**CI (95%)**	**Haz. Ratio**	**CI (95%)**
Rural in-migrant	1.08	0.82-1.43	1.06	0.80-1.41	1.08	0.81-1.43
Urban in-migrant	0.69**	0.51-0.92	0.68**	0.51-0.91	0.71*	0.53-0.95
Sex of child (M)			0.96	0.82-1.11	0.96	0.83-1.12
Teenage mother			1.07	0.89-1.29	1.05	0.87-1.27
Post-primary education					0.56**	0.41-0.75
Thatch and mud house					0.94	0.73-1.20
Mabati (zinc) roof					0.88	0.73-1.10
Untreated water					1.23**	1.05-1.43
Salaried employee					0.65	0.41-1.02

Note: *=significant 95 percent CI; ** significant at 99 percent CI. No of Deaths = 818. No of Children = 6,658. No of Observations = 38,210. Log Chi Squared = 104.11. Degrees of Freedom = 5.

References

Adazu, K., Lindblade, K.A., Rosen, D.H., Odhiambo, F., Ofware, P., Kwach, J., van Eijk, A.M., Decock, K.M., Amornkul, P., Karanja, D., Vulule, J.M. and Slutsker, L. 2005. Health and demographic surveillance rural western Kenya: A platform for evaluating interventions to reduce morbidity and mortality from infectious diseases. *American Journal for Tropical Medicine and Hygiene*, 73(6), 1151–58.

APHRC. 2002. *Population and Health Dynamics in Nairobi Informal Settlements*. Nairobi: APHRC.

Barrera, A. 1990. The role of maternal schooling and its interaction with public health programs in child health production. *Journal of Development Economics*, 32(1), 69–91.

Bicego, G.T. and Boerma, J.T. 1993. Maternal education and child survival: a comparative survey data from 17 countries. *Social Science and Medicine*, 36(9), 1207–27.

Blakely, T., Atkinson, J., Kiro, C. and d'Souza, A. 2003. Child mortality, socioeconomic position and one-parent families: independent associations and variation by age and cause of death. *International Journal of Epidemiology*, 32(3), 410–18.

Brockerhoff, M. 1990. Rural-to-urban migration and child survival in Senegal. *Demography*, 27(4), 601–16.

Brockerhoff, M. 1994. The impact of rural-urban migration on child survival. *Health Transition Review*, 4, 127–49.

Brockerhoff, M. 1995. Child survival in big cities: the disadvantage of migrants. *Social Science and Medicine*, 40(10), 1371–83.

Caldwell, J.C. *The Sahelian Drought and Its Demographic Implications*. Washington DC: American Council on Education.

Central Bureau Statistics (CBS) Kenya 1990. *Kenya Demographic and Health Survey 1989*. Kenya: Central Bureau Statistics.

Central Bureau Statistics (CBS) Kenya 1999. *Kenya Demographic and Health Survey 1998*. Kenya: Central Bureau Statistics.

Central Bureau Statistics (CBS) Kenya 2004. *Kenya Demographic and Health Survey 2003*. Kenya: Central Bureau Statistics.

Claeson, M. and Waldman, R. 2000. The evolution of child health programmes in developing countries: from targeting diseases to targeting people. *Bulletin of the World Health Organization*, 78(10), 1234–45.

Clark, S.J., Collinson, M.A., Kahn, K., Drullinger, K. and Tollman, S.M. 2007. Returning home to die: Circular labour migration and mortality in South Africa. *Scandinavian Journal of Public Health*, 35(3), 35–44.

Cleland, J.G. and Ginneken, J. 1988. Maternal education and child survival in developing countries: the search for pathways of influence. *Social Science and Medicine*, 27(12), 1357–68.

Colvin, L.G. 1981. Senegal, in *The Uprooted of the Western Sahel: Migrants' Quest for Cash in the Senegambia*, edited by L.G. Colvin. New York: Praeger

Coreil, J. and Genece, E. 1988. Adoption of oral rehydration therapy among Haitian mothers. *Social Science and Medicine*, 27(1), 87–96.

Crump, J., Otieno, P.O., Slutsker, L., Keswich, B.H., Rosen, D.H., Hoekstra, R.M., Vulule, J.M. and Luby, S.P. 2005. Household based treatment of drinking water with flocculent-disinfectant for preventing diarrhoea in areas with turbid source water in rural western Kenya: cluster randomized controlled trial. *British Medical Journal*, doi:10.1136/bmj.38512.61868.EO (published 26 July 2005).

Debpur, W.P., Akazili, J. and Nyarko, P. 2005. Health inequities in the Kassena-Nankana district in of northern Ghana, in *Measuring Health Equity in Small Areas Health Findings from Demographic Surveillance Systems*. Indepth Network: 45–65.

Defo, B.K. 1996. Areal and socioeconomic differentials in infant and child mortality in Cameroon. *Social Science and Medicine*, 42(3), 399–420.

Desai, S. and Alva, S. 1998. Maternal education and child health: is there a strong relationship? *Demography*, 35(1), 71–81.

Galvao, C.E., da Silva, A.A., dos Reis Filho, S.A., Novochadlo, M.A. and Campos, G.J. 1994. Oral rehydration therapy for acute diarrhoea in a region of north-eastern Brazil, 1986–1989 (in Portuguese). *Rev Saudi Publicia*, 28(6), 231–47.

Guest, P. 1998. Assessing the consequences of internal migration: methodological issues and a case study on Thailand, in *Migration, Urbanization and Development: New Directions and Issues*, edited by R.E. Bilsborrow. New York: United Nations Population Fund and Kluwer Academic Publishers.

Guilkey, D. and Riphahn, R. 1998. The determinants of child mortality in the Philippines: estimation of a structural model. *Journal of Development Economics*, 56(2), 281–305.

Hoque, B.A., Chakraborty, J., Chowdhury, J.T., Chowdhury, U.K., Ali, M., el Ariefeen, S. and Sack, R.B. 1999. Effects of environmental factors on child survival in Bangladesh: a case control study. *Public Health*, 113(2), 57–64.

Islam, M.M. and Azad, K.M.A.K. 2008. Rural-urban migration and child survival in Urab Bangladesh: are the urban migrants and poor disadvantage? *Journal of Biosocial Science*, 40, 83–96.

Kevin, J. and Thomas, A. 2007. Child mortality and socioeconomic status: an examination of differentials by migration status in South Africa. *International Migration Review*, 41(1), 40–74.

Khe, N.D., Dung, P.H., Phuc, H.D., Minh, H.V., Thanh, N.X., Ericson, B., Diwan, V. and Chuc, T.H.K. 2005. Health and healthcare: equity aspects in Filabavi, Vietnam, in *Measuring Health Equity in Small Areas Health Findings from Demographic Surveillance Systems*. Indepth Network: 45–65 Review, 5, 143–63.

Kiros, G. and White, M. 2004. Migration, community context and child immunization in Ethiopia. *Social Science and Medicine*, 59(12), 2603–16

Madise, N.J. and Diamond, I. 1995. Determinants of infant mortality in Malawi: an analysis to control for death clustering within families. *Journal of Biosocial Sciences*, 27(1), 95–106.

Marcos, J. and Dacunha, P. 1998. New trends in urban settlement and the role of intra-urban migration: the case of Sao Paulo/Brazil, in *Migration, Urbanization and Development: New Directions and Issues*, edited by R.E. Bilsborrow. New York: United Nations Population Fund and Kluwer Academic Publishers.

McElroy, P.D., ter Kuile, F.O., Hightower, A.W., Awhile, W.A., Phillips-Howard, P.A., Oloo, A.J., Lal, A.A. and Nahlen, B.L. 2001. All-cause mortality among young children in western Kenya VI: the Asembo Bay cohort project. *American Journal of Tropical Medicine and Hygiene*, 64(1–2 Suppl), 18–27.

Pebley, A.R., Goldman, N. and Rodriquez, G. 1996. Prenatal and delivery care and childhood immunization in Guatemala: do family and community matter? *Demography*, 33, 2312–47.

Pongou, R., Ezzati, M. and Salomón, S.A. 2006. Household and community socioeconomic and environmental determinants of child nutritional status in Cameroon. *Bio Med Central Public Health*, 6(98).

Rutstein, S. 2000. Factor associated with trends in infant and child mortality in developing countries during the 1990s. *Bulletin of The World Health Organization*, 78(10), 1256–70.

Sastry, N. 1997a. Family-level clustering of childhood mortality risk in Northeast Brazil. *Population Studies*, 51(3), 245–61.

Sastry, N. 1997b. What explains rural-urban differentials in child mortality in Brazil? *Social Science and Medicine*, 44(7), 989–1002.

Shears, P. and Lusty, T. 1987. Communicable disease epidemiology following migration: studies from the African famine. *International Migration Review*, 21(3), 783–95.

Spencer, H.C., Kaseje, D.C., Moseley, W.H., Sempebwa, E.K., Huong, A.Y. and Roberts, J.M. 1987. Impact on mortality and fertility of a community-based malaria control program in Saradid, Kenya. *Annals of Tropical Medicine and Parasitology*, 81(supl 1), 36–45.

Ssengonzi, R., De Jong, G.F. and Stokes, C.S. 2002. The effect of female migration on infant and child survival in Uganda. *Population Research and Policy Review*, 21(5), 403–31.

Stephenson, R., Matthews, Z. and McDonald, J.W. 2003. The impact of rural-urban migration on under-two mortality in India. *Journal of Biosocial Science*, 35(1), 15–31.

Thomas, D., Strauss, J. and Henriques, M. 1990. Child survival, height for age and household characteristics in Brazil. *Journal of Development Economics*, 33, 197–234.

Thomas, K.J.A. 2007. Child mortality and socioeconomic status: an examination of differentials by migration status in South Africa. *International Migration Review*, 41(1), 40–74.

Wolfe, B. and Behrman, J. 1982. Determinants of child mortality, health and nutrition in a developing country. *Journal of Development Economics*, 11, 163–93.

Zhu, J. 1998. Rural out-migration in China: a multilevel model, in *Migration, Urbanization and Development: New Directions and Issues*, edited by R.E. Bilsborrow. New York: United Nations Population Fund and Kluwer Academic Publishers.

Chapter 10

Migration and Adult Mortality in Rural Southern Mozambique: Evidence from the Demographic Surveillance System in Manhiça District

Ariel Nhacolo, Delino Nhalungo, Charfudin Sacoor,
Leonildo Matsinhe, John Aponte and Pedro Alonso

Background

Trends in adult migration in rural southern Mozambique indicate that the flow still remains mainly towards Maputo City and South Africa and in some cases Swaziland. This migration pattern, which dates back to the nineteenth century and continues to grow, has been reasonably analysed, but little is known about the mortality of migrants. Studying migrant's mortality is important for Mozambique since labour migration, especially to South Africa, is reported to be associated with increased adult mortality related to tuberculosis, HIV/AIDS and silicosis (Kahn et al. 1999, Jochelson et al. 1991, Lurie 2000). The all-cause adult mortality in Mozambique has been sharply increasing, with the probability of dying between the ages of 15–60 (${}_{45}q_{15}$) rising from 0.1 in 1980 to 0.6 in 2000 in Mozambique and from 0.4 in 1998 to 0.6 in 2005 in the Manhiça Demographic Surveillance System (DSS) (Nhacolo et al. 2006). Bradshaw and Timaeus (2006) presented ${}_{45}q_{15}$ of 0.6 among men in Mozambique, Lesotho and Namibia and of 0.7 in Botswana and Malawi during 2002.

Although data on reasons for migration are not collected in the Manhiça HDSS, it is known that migrants are leaving the study site for education and employment. This fits with other literature from southern Africa that documents high levels of labour migration (Preston-Whyte et al. 2006, Collinson et al. 2006), especially as a strategy for obtaining livelihoods for rural households. The labour migrants are mostly men travelling alone and after retirement or when sick they are highly likely to return home to convalesce or die (Clark et al. 2007). This study first examines the proposition that migrants who leave Manhiça to go and work elsewhere have lower mortality, because they are younger and are healthy people capable of working. Furthermore, the study examines the hypothesis that living outside the study site increases the risk of contracting diseases (for example HIV) in the destination place, which results in a higher mortality amongst returning migrants.

Aim of the Study

The overall aim is to examine the relationship between migration and adult mortality in rural southern Mozambique between 1998 and 2005. This includes the following questions: is migration (in and out of the HDSS area, including return migration) associated with higher adult mortality? What are the factors associated with mortality in migrants? How soon after the return of the migrant does death usually occur?

The hypotheses tested here are:

1. In general, adult migrants experience a lower mortality due to a positive selection effect, because people who are healthier may be more prone to migration (the healthy migrant hypothesis).
2. Due to HIV and other causes of death, the healthy migrant effect may be diminished due to increasing mortality of return in-migrants.

Methods

The data for this study come from the HDSS that is being maintained since 1996 by the Manhiça Health Research Centre (CISM) in Manhiça district, Maputo Province, Mozambique. A resident is defined as any person who lives in the study area and expects to stay for at least the next three months. During the surveillance every household is visited at least twice a year and all vital events, including births, deaths and migrations are recorded. The semestral visits are complemented by weekly updates by the key informants in the community and daily hospital visits, to avoid possible omissions of some events if only semestral visits were used. All the vital events are linked to the individual by the Permanent Identification Number, which is uniquely issued for each resident. This allows for an accurate follow-up avoiding duplication of individuals in the database and most importantly, allows for estimates of accurate rates by dividing the number of events by the number of person years of observation (Nhacolo et al. 2006).

Migration occurs when a person enters or leaves the study area or changes residence within the area, for at least three months or with intention to do so (Alonso et al. 2002). But for the purpose of this study a migrant is anyone who has ever moved in or out of the HDSS area and a non-migrant is someone who has never left the study area, irrespective of having moved internally.

A person quarter file was constructed that contains one record for each three-months period lived by each individual in the study population. The first person quarter was started on the date of entry into the database, either through in-migration or enumeration in the baseline census. The values of the attributes were defined at the beginning of each person quarter, including sex, date of birth, age, education, death, date of death, out-migration, in-migration as well as duration since last in-migration.

Migration rates were calculated by dividing the number of person quarters that started as a migrant by the number of person quarters experienced by the population. The same procedure was used to calculate mortality rates. For testing the hypotheses on migration and death an event history analysis was used to estimate the quarterly hazard of death (logistic regression)[1] as a function of whether or not the person in-migrated in the observation period. The analysis was clustered at the Permanent Identification Number to remove the independence between multiple records of the same individual, which underlies the logistic regression method.

The phenomenon of returning home to die is examined further in a multivariate analysis, examining what factors are associated with the death of an adult in-migrant. The variables explored are sex, education, duration since last in-migration, study period, frequency of migration and the place where the person was coming from at last entry in the study area. The study is restricted to individuals at working ages (20–59 years old) due to the fact that the relationship between adult labour migration and mortality will be studied. Education is modelled as a dichotomous variable where one (1) represents the people who attended some secondary school or higher education and zero (0) those who completed only primary school or less. Duration is grouped in four categories: migrated within the previous quarter; four to six months ago; seven to nine months ago; and ten months or more ago. The study periods compared are 1998–2001 and 2002–2005. The frequency of migration (both out and in-migration) is grouped in four categories: migrated only once; two times; three times; four times or more. The place of origin at last entry date was grouped in three categories (South Africa; Maputo City and other). These factors are examined in the same logistic model with 'death-ended' person quarters modelled as a function of all person quarters that were experienced in the population at risk.

Results

From January 1998 to December 2005, 46,778 adults aged 20–59 were followed (736,433 person quarters). Around 58 percent (26,929) of these adults were migrants (have moved in or out of the HDSS area at least once during the study period). About four percent (1,095) of the migrants have died; 14,787 (55 percent) were alive and 11,047 (41 percent) exited the HDSS area (unknown survival

1 Event History Analysis is a prediction of an outcome from the sequence of previous exposures, in this case the exposure is migration and the outcome is death. The question to answer is: does the exposure change the probability of occurrence of the outcome?

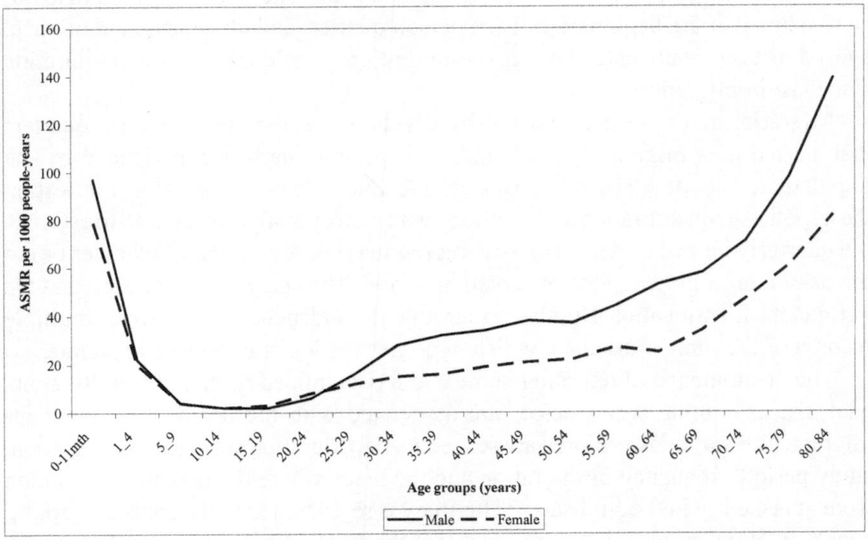

Figure 10.1 Age-specific mortality rates by sex, Manhiça HDSS area 1998–2005

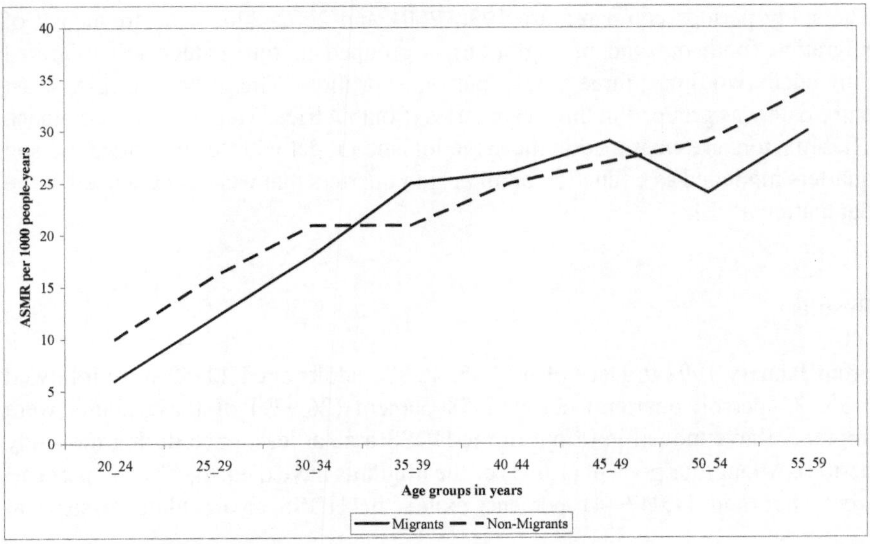

Figure 10.2 Age-specific mortality rates by migrant status, Manhiça HDSS area 1998–2005

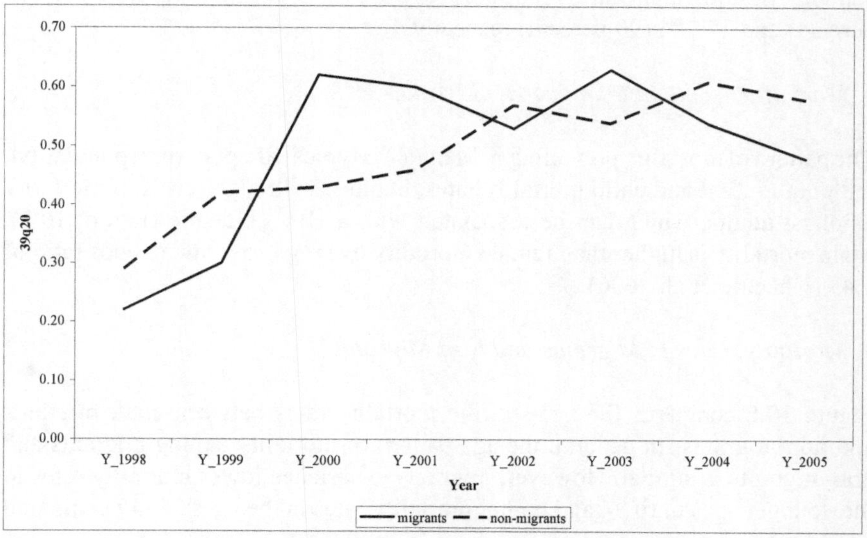

Figure 10.3 39q20 from 1998–2005 by migrant status, Manhiça HDSS area

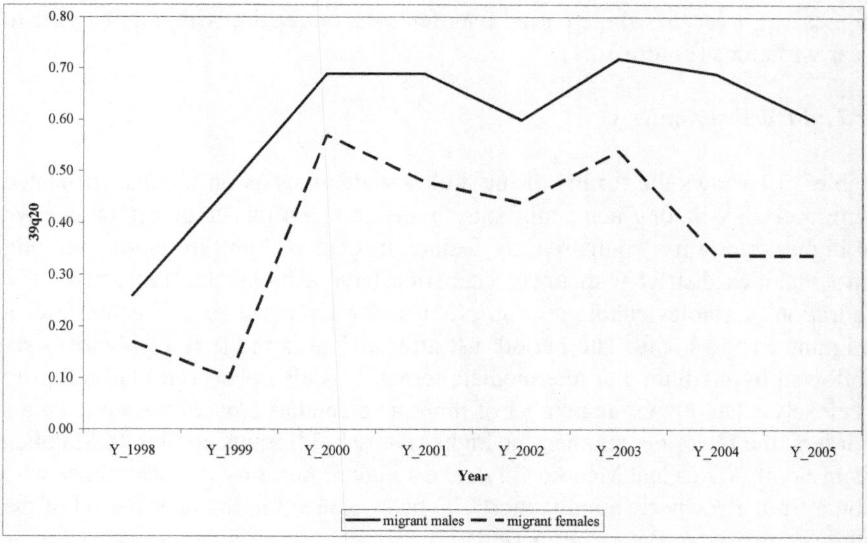

Figure 10.4 39q20 for migrants from 1998–2005 by sex, Manhiça HDSS area

status). The equivalent survival figures for the 19,849 on-migrants are 2,065 (10 percent) and 17,784 (90 percent) respectively.

1. Mortality Profile in Manhiça HDSS Area

The pattern of mortality prevailing in Manhiça is typical of a poor rural population, with high infant and child mortality rates, but also with high levels of mortality amongst adults, which can be associated with a HIV epidemic (Figure 10.1). Male mortality is higher than female mortality over age 25, with an odds ratio of 1.43 (Nhacolo et al. 2006).

2. Mortality Rates in Migrants and Non-Migrants

Figure 10.2 compares the age-specific mortality rates between adult migrants and non-migrants. In general, the age pattern of mortality among migrants and non-migrants is similar. However, migrants experience lower mortality rates in the younger ages until 34 and higher mortality rates in the ages 35–49 compared to non-migrants.

Figure 10.3 gives an indication of changes in adult mortality over time in Manhiça among migrants and non-migrants. The general trend is that there is an increase of the probability of dying between the ages of 20 and 59 ($_{39}q_{20}$) among non-migrants and migrants, although there have been year to year variations.

Splitting the migrant's $_{39}q_{20}$ by sex shows that the male migrants have a higher $_{39}q_{20}$ over the whole period than females, but again, with notable year to year variations (Figure 10.4).

3. Multivariate Analysis

Table 10.1 shows the results of the multivariate analysis on factors associated with mortality among adult migrants. It shows again that male migrants have a higher mortality compared to female migrants. Adult migrants moving into Manhiça district with lower education have a higher mortality risk. The 'duration' variables enable a description of the temporal risk of death after a migrant arrives home. The period just after arrival is the time of highest risk, followed by medium and intermediate terms. It is also observed that mortality is closely related with the number of times a person has crossed the study area's borders: the fewer the number, the higher the risk. Migrants coming to Manhiça from South Africa and Maputo City have a higher mortality risk than those who come from elsewhere. Finally, mortality has worsened in the later period of the study compared to the earlier period.

Table 10.1 Factors associated with mortality among adult migrants, Manhiça 1998–2005 (multivariate analysis)

Factors associated with mortality	Odds ratio	P- value	95% CI
Male migrant compared to female	1.47	0.00	1.30–1.65
Younger migrants (20–34 years old) compared to older migrants (35–59 years)	0.46	0.00	0.41–0.52
Lower education compared to higher education	1.69	0.00	1.34–2.14
Short-term in-migrant (within 3 months) compared to long-term immigration (10+ months)	8.30	0.00	6.68–10.30
Medium-term in-migrant (4–6 months) compared to long-term immigration (10+ months)	7.88	0.00	6.42–9.68
Intermediate-term in-migrant (7–9 months) compared to long-term immigration (10+ months)	3.90	0.00	3.06–4.96
Migrated once compared to migrated 4+ times	16.14	0.00	10.40–25.04
Migrated two times compared to migrated 4+ times	3.97	0.00	2.61–6.04
Migrated three times compared to migrate 4+ times	2.76	0.00	1.75–4.34
Last place of origin RSA compared to other places of origins	1.96	0.00	1.64–2.34
Last place of origin Maputo City compared to other places of origins	1.71	0.00	1.48–1.98
Later study period compared to earlier (2002–2005 vs 1998–2001)	1.75	0.00	1.50–2.06

Discussion

Immigration of citizens of Mozambique, Botswana, Lesotho and Swaziland to the South African mines is the best known case of labour migration in sub-Saharan Africa (Oucho 1998). The relation between migration and health is complex and operates both ways, from migration to health and from health to migration and can be either positive or negative (Garenne 2006). The economic benefits to the household of sending a migrant to the city or neighbouring country have been documented (Taylor 1999). Studies also show that migrants are positively selected and that better health goes together with a higher chance of migration. Once people have migrated they can bring the health knowledge and technology back with them, which improves their own outcomes but can also have a wider effect in the household and community. In this respect migration is an agent of social change and improved perspectives on health care seeking, nutrition and the value of education (Kuhn 2006, Lindstorm and Munoz-Franco 2006). On the other hand, migrants may encounter the negative effects of exposure to HIV infection which is particular to the experience of labour migration at the start of the new century in Southern Africa.

First this chapter examines whether migration has a positive or negative influence on mortality in the 1998–2005 period. There is a modest positive relationship between migration and survival in the age group 20–34. This is likely due to a selection effect, whereby the most healthy members of the population are also the most likely to migrate. Between ages 35 and 49 the pattern changes and a significantly higher risk of mortality exists for in-migrants, which can be associated with returning home to die.

Regarding duration this chapter finds a similar effect as the Clark et al. (2007) paper, namely that mortality is higher in the migrants who only recently came back, indicating that these returned home to die. The mortality rate is lower in those who have been back for more than a year, as they may come to Manhiça for other reasons than illness, such as family reasons and employment.

The education variable reflects whether or not a person has experienced some secondary level of education. The effect on mortality is significantly protective when controlling for the other variables. It is likely that this represents a selection effect, whereby the people most likely to get some secondary education have socio-economic advantages and may have skills, experiences or values that protects them against mortality, such as health seeking behaviour and nutrition.

Because a person needs to have at least three months of residence before he or she is recorded into the database (or with a declared intention to stay), it is believed that the HDSS fieldworkers may miss some people who have been away, but came back and died immediately. This situation requires a careful clarification from respondents to the fieldworkers on why the person has come to Manhiça, to decide whether or not to register the in-migration and the death. The increased mortality risk of people with one migration reflects not only returning home to die, but also the fact that people who migrated regularly come back to Manhiça for a short visit (less than three months), but are only being registered at the last entry. There is no way of tracking HDSS members when they are away, unless they come back alive. Therefore it is believed that part of the 41 percent of migrants who have exited may have died within the HDSS area, but are not recorded. This implies that the effects seen in the chapter are a minimum estimation of returning home to die and that the actual figures are likely to be higher.

Conclusions

This study provides findings that are consistent with the hypothesis that there is a healthy migrant effect at younger ages, but due to HIV and other causes of death this effect may be diminishing with age. For the adult migrants, the shorter the duration since last in-migration, the more likely the death of the in-migrant, indicating that they came home to die. The evidence suggests that increasing numbers of circular migrants of prime working age are becoming ill as they age in the urban areas where they work and come home to be cared for and to die in the rural areas where their families live. This shifts the burden of caring for them in their terminal illness to their families and the rural healthcare system with significant consequences for the distribution and allocation of healthcare resources.

References

Alonso, P.L., Saute, F., Aponte, J.J., Gomez-Olivé, F.X., Nhacolo, A., Thompson, R., Macete, E., Abacassamo, F., Ventura, P.J., Bosch, X., Menendez, C. and Dgedge, M. 2002. Population and health in developing countries, in *Populations and Their Health In Developing Countries, Volume 1: Population, Health and Survival At INDEPTH Sites.* Ottawa: International Development Research Centre (IDRC) Press.

Bradshaw, D. and Timaeus, I. 2006. Levels and trends of adult mortality, in *Disease and Mortality In Sub-Saharan Africa*, edited by D.T. Jamison, R.G. Feachem, M.W. Makgoba, E.R. Bos, F.K. Baingana, K.J. Hofman and K.O. Rogo. Second edition. The World Bank. Available at: www.ncbi.nlm.nih.gov/bookshelf/ (last accessed: 7 January 2009).

Clark, S.J., Collinson, M.A., Kahn, K. and Tollman, S.M. 2007. Returning home to die: urban to rural migration and mortality in rural South Africa. *Scandinavian Journal of Public Health*, 35(Suppl. 69), 35–44.

Collinson, M.A., Tollman, S.M., Kahn, K., Clark, S.J. and Garenne, M. 2006. Highly prevalent circular migration: households, mobility and economic status in rural South Africa, in *Africa on The Move: African Migration and Urbanisation In Comparative Perspective*, edited by M. Tienda, S.E. Findley, S.M. Tollman and E. Preston-Whyte. Johannesburg: Wits University Press.

Garenne, M. 2006. Migration, Urbanisation and Child Health in Africa: a global perspective. *Africa on The Move: African Migration and Urbanisation In Comparative Perspective.* M. Tienda, S.E. Findley, S.M. Tollman and E. Preston-Whyte. Johannesburg: Wits University Press.

Jochelson, K., Mothibeli, M. and Leger, J.P. 1991. Human Immunodeficiency Virus and migrant labour in South Africa. *International Journal of Health Services*, 21(1), 157–73.

Kahn, K., Tollman, S., Garenne, M. and Gear, J. 1999. Who dies from what? Determining cause of death in South Africa's rural north-east. *Tropical Medicine and International Health*, 4(6), 433–41.

Kuhn, R. 2006. The effect of father's and siblings' migration on children's pace of schooling in rural Bangladesh. *Asian Population Studies*, 2(1), 69–92.

Lindstorm, D.P. and Munoz-Franco, E. 2006. Migration and maternal health services utilization in rural Guatemala. *Social Science and Medicine*, 63(3), 706–21.

Lurie, M. 2000. Migration and AIDS in southern Africa: a review. *South African Journal of Science*, 96, 343–47.

Nhacolo, A., Nhalungo, D.A., Sacoor, C.N., Aponte, J.J., Thompson, R. and Alonso, P. 2006. Levels and trends of demographic indices in Southern Rural Mozambique: Evidence from Demographic Surveillance System in Manhiça District. *Bio Med Central Public Health*, 6, 29.

Oucho, J. 1998. Regional integration and labour mobility in eastern and southern Africa, in *Emigration Dynamics in Developing Countries Volume 1: Sub-Saharan Africa*, edited by R. Appleyard. Aldershot: UNFPA, IOM Ashgate.

Preston-Whyte, E., Tollman, S.M., Landau, L. and Findley, S.E. 2006. African Migration in the Twenty-First Century, in *African On The Move: Migration in Comparative Perspective*, edited by M. Tienda, S.E. Findley, S.M. Tollman and E. Preston-Whyte. Johannesburg: Wits University Press.

Taylor, J.E. 1999. The new economics of labour migration and the role of remittances in the migration process. *International Migration*, 37(1), 63–88.

Chapter 11
Migration and Under Five Morbidity in Bavi, Vietnam

Ho Dang Phuc, Nguyen Xuan Thanh and Nguyen Thi Kim Chuc

Introduction

The population boom experienced in Vietnam during the 1970s to the early 1990s with a growth rate of about two percent per year remarkably influenced the living conditions in rural regions. Each year the Vietnamese population was growing by 1.2 million on average, resulting in about 900,000 to 1.2 million young people entering the labour force every year during 1979–1999 (Development Analysis Network 2001). The share of the working age population (15–60 years old) within the total population increased from 47 percent in 1979 to 52.6 percent in 1989 to 57.5 percent in 1999. In rural areas, especially in the Red River Delta, the population boom made land scarce and human resources more abundant. This resulted in un- and underemployment of a large part of the population. According to the Statistical Yearbook 2003, the population density of the Red River Delta was about 12 persons per hectare of natural land and 20 persons per hectare of agriculture land in 2002 (General Statistics Office 2003). The serious land shortage and unemployment resulted in low living standards for the population in the countryside and strongly pressed migration from the rural regions to urban areas where people may have a better opportunity to get a job (World Health Organisation 2006).

On the other hand, the economic reform which was introduced in 1986, including more liberal policies (Doi Moi), has crucially transformed Vietnam's political, economic and social contexts. This let the country enter into a new stage of development; the stage of industrialization and modernization. In the process of industrialization, a number of enterprises rose up around big cities, several industrial/processing zones were created all over the country which became dynamic economic centres with many firms financed with local and foreign money operating in it. Those zones have absorbed millions of workers and indirect workers (those who do not work in the firms but have businesses linked to the enterprises or supplying services to the workers). The labour army came from different areas and resulted in a rapid urbanization. In the meantime, in the countryside the introduction of Doi Moi, with a market-oriented economy and with de-collectivization and the implementation of the household contract system in agriculture, had released the rural workforce from their tie to the land

and prompted them to leave the homeland and go to other places with a more convenient livelihood.

As a consequence the migration has been steadily increasing year after year in Vietnam for the last two decades. Recently some studies have been conducted on the migration phenomenon in Vietnam. Loi attempted to analyse the 'pull and push factors' influencing the rural-to-urban migration trend (Cu Chi Loi 2009). They showed that the poverty yielded by land shortage and unemployment in rural areas was a 'pushing factor'. On the other hand, the rapid industrial development implying labour need in urban areas triggered the attraction of migrants from the countryside ('pulling factor'). The author also described the migration flows from different rural regions as well as the structure of the employment of migrants in their new settlements at industrial zones. Winkels and Adger (2002) explained the importance of social capital in the form of migrant networks for long distance rural-to-rural migration (from northern delta regions to southern mountain areas). Another study by Anh et al. (2003) gave an overview of trends, patterns and policy implications of overall migration in the whole country, including domestic and overseas migration.

The above mentioned studies mainly analysed the socio-economic patterns of migrant's communities in the new settlements. The impact of migration on the home communities of the migrants was not really addressed. Only the study by Anh included a conclusion on the economic and social role of remittances from international labour migrants, other effects of migration movements were not incorporated. Therefore the changes in rural societies occurring as a result of all kinds of migration activities would be an interesting topic for further studies.

Among the migrants from rural areas, a large part has strong connections with relatives left at their residency in the homeland. Taking on different kinds of jobs, as workers in enterprises, as street vendors in big cities or housekeepers for urban families, they share a part of their income earned with their family to help improving the living standards of those left behind. However, the migration process can also result in several social problems at both ends of the movement flow. For instance, in the countryside the absence of young adults who leave the household for a long time could result in children less taken care of, who become sick easier and so on.

Using longitudinal data collected between 1999 and 2005 at Filabavi, a demographic field laboratory situated in a rural district in North Vietnam, this study attempts to describe migration patterns in Bavi district in that period and investigate the relationship between migration of parents and morbidity of their under five children.

Methodology

Setting

Bavi is a rural district of Ha Tay province in northern Vietnam, 60 km west of Hanoi (the capital city). The district consists of 32 communities, covers an area of 410 km^2, including lowland, highland and mountain zones and ranges in altitude from 20 to 1297 metres above sea level. An average of 532 m^2 is used for agriculture per capita. The climate is typical for northern Vietnam, with a monsoon tropical climate: a wet season from May to October with high temperatures, heavy rainfall and storms and a dry season from November to April with generally lower temperatures but hot, dry winds for about a month before the rains start.

The district population is approximately 240,000 (estimated in 1999) of which 91 percent belongs to the Kinh ethnic (majority) group, eight percent to the Muong and there are minorities of Dao, Thai, Tay, Hoa and Khme. Illiteracy is low, at only about 0.4 percent of the adult population (>15 years of age). About 69 percent of the adult population has completed primary school, 21 percent lower secondary level, nine percent higher secondary level and 0.5 percent higher education.

The main economic activities in the district are farming and livestock breeding (81 percent) with the most important products being wet rice, cassava, corn, soybean, green beans and some fruits such as bananas, pineapples, mandarins, papayas and lychees. Traditionally there are two main rice crops in June and November annually. Other economic activities are forestry (eight percent), handicraft (six percent), small trade (three percent), fishing (one percent) and transport (one percent). In 1999 the average annual income was 150 USD; in 2005 it was 275 USD.

In the area there is one district hospital (DH) and there are three policlinics and 32 community health stations (CHS). The distance from a household to a CHS was estimated to be 1.1 km on average in the lowland, 1.6 km in the highland and 2.3 km in the mountainous area. The average distance from a community to the DH was estimated to be about 5.2 km in all areas, with the longest distance being 12.7 km for one community in the mountains. Distance to the provincial hospital was estimated to be about 15 km on average with the longest distance being 25 km for remote communities.

The study was conducted within a population-based demographic surveillance site, called Filabavi, situated in Bavi district. A random sampling of villages was performed, with a probability proportional to the population size in each unit. The Filabavi includes 67 clusters randomly selected from a total of 352 clusters, comprising 11,929 households with 49,447 inhabitants at 1 July 1999, which is approximately 20 percent of the total population in the district. Socio-economic information of the households and their members were collected at the Household Baseline Survey completed at the beginning of 1999, which was followed by Census Surveys conducted every two years (2001, 2003 and 2005). Following the Baseline Survey, all households were visited every three months to obtain information about vital events, primarily birth, death and migration, as well as information on pregnancy, change of marital status, self reported sickness episodes and so on. The household interviews were conducted by 36 female, specifically trained and full-time employed interviewers. Five percent of the households (randomly selected) were re-interviewed by six surveyors for quality control purposes.

Definition of Variables

In the quarterly visits of the households in the sample, interviewers asked the household head about migration and sickness episodes that had occurred with respect to household members since the last visit. An in-migration event was recorded when during the period a person had come to live with the household and intended to live there for more than three months. An out-migration event was recorded if since the last visit a person had left this household and intended to stay away for more than three months. In this study only external migration events were dealt with, that means that all movements within the surveillance site (internal migration events) were not included in the analysis.

A sickness episode (any symptom of cough, fever, headache, respire, colic, digestion, bone ache, injury, blood hypertension and heart disease) that had occurred in the period of four weeks before the interview was reported at a quarterly visit in the following situations:

- The sickness started before the four week period and the ill person had already recovered at the time of interview, but it covered most of the period;
- The sickness started before the four week period and continued still at the time of interview;
- The sickness started within the four week period and the ill person had already recovered at the time of interview;
- The sickness started within the four week period and continued still at the time of interview.

Regarding socio-economic information, economic status of the households as well as education and occupation of all members were recorded at the Baseline Survey in 1999 and the Census Surveys in 2001, 2003 and 2005. A person was said to have 'Primary or less' education if at the moment of the review he was illiterate or had passed only up to class five in the 12 year school system; 'Lower secondary education' if he finished class six up to nine; 'Higher secondary school or over' education if he passed class ten up to twelve, finished professional school or university.

There are four occupational groups: 'Farmer'; 'Government staff' (persons working in public service facilities); 'Worker or handicraft maker' and 'Other' which contains the rest, for example small traders, housewives, tailors, drivers, contractors, hired workers and so on, but also retired, elder or jobless people.

To classify households into different economic groups, a variable called 'wealth index' was constructed using Principal Component Analysis (PCA) (Gwatkin et al. 2000, Houweling et al. 2003). The wealth index is the first principal component of selected economic, housing, land area and household asset variables reported at Baseline or Census Surveys. The study households were then divided into five 'wealth quintiles', that is, groups defined by the quintiles of the wealth index and named 'Poorest', 'Poor', 'Average', 'Upper average' and 'Richest', respectively.

Economic status of a person was defined by the wealth quintile of the household. Children were divided into 'Poor' and 'Non-poor' groups depending on whether the mother was classified into the first two quintiles or into the last three quintiles.

Design, Sampling and Sample Size

Because of the quarterly surveys, each person has been observed several times, including the migration and sickness episodes. So the 'sample size' of the study must be understood as the total number of person observations cumulated through all interviews in the period from 1999–2005.

Socio-economic information (education, occupation and economic status) of any person was also revised every two years and could change over time. The link between socio-economic information and personal observations was made depending on the date of the given visit and the most recent census. For example, all the observations with visit date before 1 July 2000 were linked to socio-economic characteristics from the Baseline Survey of 1999; the observations obtained from 1 July 2000–1 July 2002 were joined with socio-economic features from the Census Survey of 2001 and so on.

Statistical Methods

Besides the PCA method used in determining the economic status of households and persons as mentioned above, the main statistical method used in this study is logistic regression. In the model for morbidity in under five children, 'sickness episode' is the dependent variable with value one (1) if a sickness episode was reported for the observed child in the given quarterly visit and with value zero (0) if no sickness episode was recorded for that child. To obtain odds ratios, all independent variables (factors) in the model were recoded into binary dummy variables, any observation belonging to the reference group of a given factor had value 0 for all dummy variables related with this factor. As all independent variables were put together into the regression equation, the odds ratios obtained represented those of groups defined by a given factor respective to the reference group of that factor, adjusting for the influence of other factors present in the model.

Results and Discussion

1. Migration Patterns

Net Migration by Year, Sex and Age The numbers of net migrants (number of in-migrants minus number of out-migrants) by year are presented in Table 11.1. In total, there were 15,602 out-migrants and 11,418 in-migrants giving -4,184 net migrants in Bavi District between June 1999 and December 2005. The number

Table 11.1 Migration by years in Filabavi

	6 – 12/ 1999	2000	2001	2002	2003	2004	2005	Total
In	1,014	1,816	1,761	1,691	1,688	1,707	1,741	11,418
Out	1,231	2,356	2,606	2,296	2,715	2,202	2,196	15,602
Net	-217	-540	-845	-605	-1,027	-495	-455	-4,184

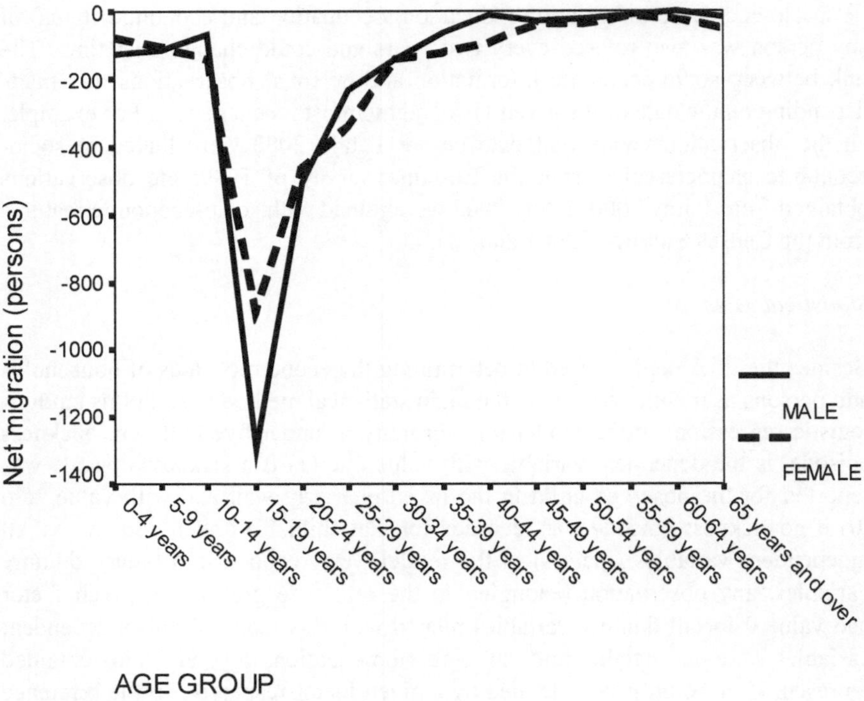

Figure 11.1 Age and net migration by gender in Filabavi

of out-migrants is larger than the number of in-migrants in all the years, with the largest difference occurring in 2003 (1,027 net migrants).

Figure 11.1 shows the numbers of net migrants by sex and age. Men seemed to out-migrate more frequently than women. Net migrants were largest in the age groups of 15–30 years for both sexes. This can be an explanation for the gaps at the ages between 20 and 35 years that can be seen in the population pyramid of the studied area in 2003 (Figure 11.2).

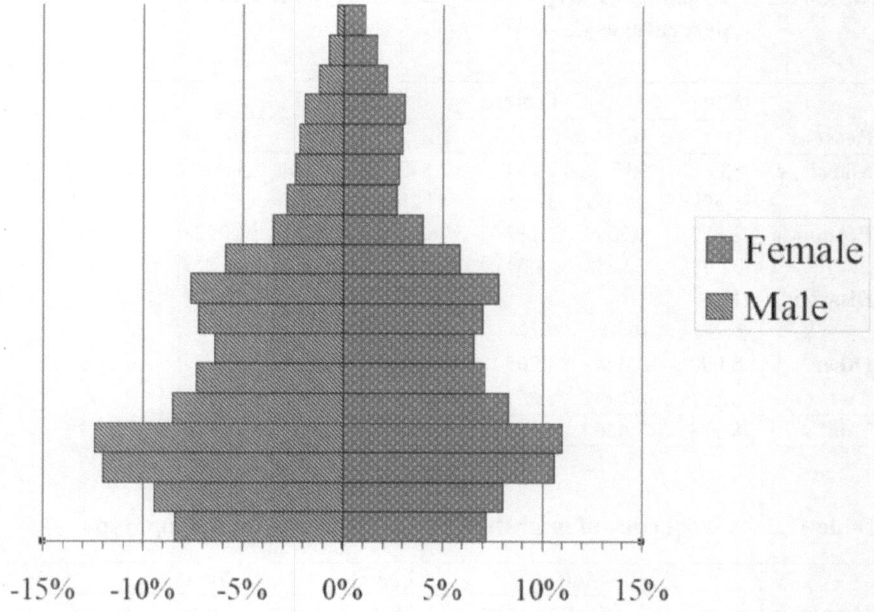

Figure 11.2 Filabavi 2003 population pyramid

Reasons for Migration by Type and Sex

The most common reason for migration was 'other' and the least was 'disaster' for both types, that is, in- and out-migration and sexes (Table 11.2). The 'other' category included people changing residence to live with someone other than their husband or wife (parents, descendants and so on), to seek health care, to study, to build or buy a new house, to sell an old house and so on. 'Economic' reasons and 'marriage' were the second and third reasons depending on types of migration for both men and women. The economic reasons were more common than marriage among out-migrants while marriage was more common among in-migrants. However, this also differed between men and women. For male migrants, the economic reasons were more common than marriage regardless of the type of migration. For female migrants, the economic reasons were more common than marriage in the case of out-migration, while for in-migration it was the other way around.

This can be explained by the fact that in rural Vietnam, men are usually the breadwinners and heads of the household. Furthermore, men are physically stronger than women, so they may find temporary or seasonal physical work (the kinds of work that farmers or unskilled people can do) easier than women do in urban areas.

Table 11.2 Reasons for migration by sex and type (numbers and percentages)

	Male		Female		Total	
Reasons	**Out**	**In**	**Out**	**In**	**Out**	**In**
Marriage	539	937	2,714	2,872	3,253	3,809
	6.5%	*16.6%*	*31.5%*	*47.2%*	*19.3%*	*32.5%*
Economic	2,680	1,257	2,117	765	4,797	2,022
	32.3%	22.2%	24.6%	*12.6%*	*28.4%*	*17.2%*
Disaster	11	16	8	14	19	30
	0.1%	*0.3%*	*0.1%*	*0.2%*	*0.1%*	*0.3%*
Other	5,058	3,444	3,765	2,430	8,823	5,874
	61.9%	*60.9%*	*43.8%*	*40.0%*	*52.2%*	*50.1%*
Total	8,288	5,654	8,604	6,081	16,892	11,735

Table 11.3 Frequency of migration by economic status, sex and type

	Male		Female		Total	
Economic status	**Out**	**In**	**Out**	**In**	**Out**	**In**
Poorest quintile	1,304	1,008	1,545	1,198	2,849	2,206
	15.7%	*16.9%*	*17.9%*	*18.7%*	*16.8%*	*17.8%*
Poor quintile	1,499	1,080	1,621	1,225	3,120	2,305
	18.0%	18.2%	18.8%	*19.1%*	*18.4%*	*18.6%*
Average quintile	1,825	1,234	1,739	1,274	3,564	2,508
	21.9%	*20.7%*	*20.2%*	*19.9%*	*21.0%*	*20.3%*
Upper-average quintile	1,790	1,176	1,779	1,229	3,569	2,405
	21.5%	*19.8%*	*20.7%*	*19.1%*	*21.1%*	*19.4%*
Richest quintile	1,899	1,450	1,931	1,492	3,830	2,942
	22.8%	*24.4%*	*22.4%*	*23.2%*	*22.6%*	*23.8%*
Total	8,317	5,948	8,615	6,418	16,932	12,366

Migration by Types, Sex and Economic Status

It is hypothesized that poor people more often migrate than rich people because they try to find better opportunities to improve their status. When reviewing Table 11.3 and 11.4 a positive association is found between migration and income. These tables show frequencies of migration and frequencies of migration for those migrating because of economic reasons, respectively, by sex, economic status and types of migration. In both tables, generally the poorest quintile migrated the least while the richest quintile migrated the most. Other chapters in this book report on this in more detail.

Table 11.4 Frequency of migration for economic reasons by economic status, sex and type

	Male		Female		Total	
Economic status	Out	In	Out	In	Out	In
Poorest quintile	314	176	299	132	613	308
	12.3%	*16.2%*	*14.7%*	*19.1%*	*13.4%*	*17.3%*
Poor quintile	459	184	386	132	845	316
	18.0%	17.0%	19.0%	*19.1%*	*18.5%*	*17.8%*
Average quintile	622	225	456	132	1,078	357
	24.4%	*20.7%*	*22.5%*	*19.1%*	*23.6%*	*20.1%*
Upper-average quintile	595	224	484	125	1,079	349
	23.4%	*20.6%*	*23.8%*	*18.1%*	*23.6%*	*19.7%*
Richest quintile	554	276	405	170	959	446
	21.8%	*25.4%*	*20.0%*	*24.6%*	*21.0%*	*25.1%*
Total	2,544	1,085	2,030	691	4,574	1,776

However, this result should be viewed in the light of the context of Bavi. Bavi is a poor rural district of Vietnam where people live on rice with an average income per month per head that was estimated to be 356,000 VND (about 23 USD) in 2005. Therefore, even people in the richest quintile may in fact not be rich; they would be considered as poor or average in other contexts where people's income and living standards are higher, such as Hanoi. Possibly, all people in the Bavi district want to improve their income by migration, but people in the richest quintile may have better conditions or better opportunities than those who are in the poorest quintile to do so. That is why the richest quintile accounted for the highest frequency of migration in Filabavi. This is supported by the results in Table 11.4, where the highest frequency of out-migration among those with economic reasons belonged to the average and upper average quintiles.

Reasons for Migration by Types and Educational Levels

Table 11.5 shows the reasons for migration by educational levels and types of migration. It is clear that for all educational levels, marriage is more frequent among in-migrants than out-migrants, while economic reasons are more frequent among out-migrants than in-migrants. For 'other' reasons, the frequency seems to decrease when the educational level increases. This is clearer in the case of in-migration where the frequency reduced from 59.2 percent among the educational level of primary education and less, to 46.2 percent among those with higher secondary school and over.

Table 11.5 Reasons for migration by educational levels and type (numbers and percentages)

	Primary and less		Lower secondary		Higher secondary and over		Total	
Reasons	**Out**	**In**	**Out**	**In**	**Out**	**In**	**Out**	**In**
Marriage	978	1,005	1,444	1,740	831	1,064	3,253	3,809
	20.7%	*28.0%*	*21.2%*	*34.6%*	*15.6%*	*34.1%*	*19.3%*	*32.5%*
Economic	974	445	2,034	965	1,788	613	4,796	2,023
	20.6%	*12.4%*	*29.8%*	*19.2%*	*33.5%*	*19.6%*	*28.4%*	*17.2%*
Disaster	7	11	10	16	2	3	19	30
	0.1%	*0.3%*	*0.1%*	*0.3%*	*0.0%*	*0.1%*	*0.1%*	*0.3%*
Other	2.762	2,122	3,339	2,313	2,722	1,440	8,823	5,875
	58.5%	*59.2%*	*48.9%*	*45.9%*	*50.9%*	*46.2%*	*52.2%*	*50.1%*
Total	4,721	3,583	6,827	5,034	5,343	3,120	16,891	11,737

Table 11.6 Frequency of migration for economic reasons by educational levels, sex and type

	Male		Female		Total	
	Out	**In**	**Out**	**In**	**Out**	**In**
Primary and less	510	218	437	178	947	396
	20.0%	20.1%	21.5%	25.8%	20.7%	22.3%
Lower secondary	973	495	955	337	1,928	832
	38.2%	45.6%	47.1%	48.8%	42.1%	46.8%
Higher secondary and over	1,063	373	637	176	1,700	549
	41.8%	34.3%	31.4%	25.5%	37.2%	30.9%
Total	2,546	1,086	2,029	691	4,575	1,777

Among those who migrated for economic reasons, people with lower secondary level education are most likely to report to have out-migrated (42 percent) or in-migrated (47 percent) (Table 11.6).

Reasons for Migration by Types and Occupations

In general, 'other' reasons were the most common among both out- and in-migrants. The 'other' reason partially took into account elderly people who migrated to live with descendants or to seek health care. This reason also deals with a part of the young people classified occupationally to the 'other' group, who changed their place of residence to study or in order to gain qualified working skills. The

Table 11.7 Reasons for migration by occupation and type (numbers and percentages)

	Farmer		Government staff		Worker and Handicraft maker		Others		Total	
	Out	In	Out	In	Out	In	Out	In	Out	In
Marriage	1,581	1,917	193	282	98	131	1,381	1,479	3,253	3,809
	23.6%	39.0%	29.7%	47.9%	18.6%	30.8%	*15.3%*	*25.5%*	*19.3%*	*32.5%*
Economic	1,786	720	213	95	213	102	2,585	1,106	4,797	2,023
	26.7%	14.6%	32.8%	16.1%	*40.3%*	*23.9%*	*28.6%*	*19.0%*	*28.4%*	*17.2%*
Disaster	6	8	0	0	0	1	13	21	19	30
	0.1%	0.2%	*0.0%*	*0.0%*	*0.0%*	*0.2%*	*1%*	*0.4%*	*0.1%*	*0.3%*
Other	3,314	2,270	243	212	217	192	5,049	3,201	8,823	5,875
	49.6%	*46.2%*	*37.4%*	*36.0%*	*41.1%*	*45.1%*	*55.9%*	*55.1%*	*52.2%*	*50.1%*
Total	6,687	4,915	649	589	528	426	9,028	5,807	16,892	11,737

Table 11.8 Frequency of migration for economic reasons by occupation, sex and type

	Male		Female		Total	
	Out	In	Out	In	Out	In
Farmer	720	311	972	310	1,692	621
	28.3%	*28.6%*	*47.9%*	*44.9%*	*37.0%*	*34.9%*
Government staff	104	54	105	38	209	92
	4.1%	5.0%	5.2%	*5.5%*	*4.6%*	*5.2%*
Worker and handicraft maker	126	64	82	31	208	95
	4.9%	*5.9%*	*4.0%*	*4.5%*	*4.5%*	*5.3%*
Others	1,596	657	871	312	2,467	969
	62.7%	*60.5%*	*42.9%*	*45.2%*	*53.9%*	*54.5%*
Total	2,546	1,086	2,030	691	4,576	1,777

migration pattern changed considerably among occupations. Among government staff, marriage was the most frequent reason for in-migration (48 percent) and the 'other' reasons became the second with 36 percent (Table 11.7). Other noticeable results is that for economic reasons, workers and handicraft makers out- migrated most often (40 percent), followed by government staff (33 percent), 'others' (29 percent) and farmers (27 percent); and percentages of out-migration are higher, almost doubled, than that of in-migration in all occupations. Frequencies of migration among those who migrated because of economic reasons by sex and occupations are shown in Table 11.8. In general, those with other occupations migrated the most, followed by farmers.

2. Morbidity of Under Five Children

To investigate the relationship between parents out-migration and morbidity of their children, a logistic model was estimated with 'sickness episode' as the dependent variable and out-migration, household economic level and occupation of the father and mother as independent variables corrected for age (as a continuous variable and age^2), sex and ethnic group of the child and education of the father and mother.

The results show that children under five years of age who have had an out-migrating mother have a 1.381 times higher risk of morbidity than children whose mother did not out-migrate. No significant difference was found between migrating and non-migrating fathers.

Regarding the influence of household economic condition on health, children under five years of age living in non-poor households have a lower possibility of getting sick than those who live in poor households (odds ratio 0.908).

Children under five years, with a father who is not a farmer, have a lower risk of morbidity, with odds ratios of 0.570 (government staff), 0.629 (worker) and 0.414 (other occupations). While the morbidity risk of children with a mother-worker is not significantly different from that of children with a mother-farmer (p-value = 0.655), children of a mother who is a government staff member or has an other occupation have lower morbidity risks, with odds ratios of 0.627 and 0.785 respectively.

When considering mothers who are government staff; they have work responsibility for only eight hours per day, so they can spend more time with their own than mothers who are farmers. Besides that they can send the child to a crèche when they are at work, while mothers who are farmers or workers usually let older children look after the younger ones. Mothers who come within the category 'other' have also more time for child care compared to mothers who are farmers or workers. That may explain the differences in morbidity risk of children.

The above results are in line with our expectations and can be explained through the level of child care and the economic situation. A child would be less likely to get sick if he/she was born in a higher income family and received more care from her/his mother.

Conclusion

This study highlighted some remarkable points in the Filabavi migration pattern. It has been pointed out that there is more out-migration than in-migration of people in Bavi District. That explains the stability of the Filabavi population, despite the fact that every year the crude birth rate is higher than the crude death rate. Men out-migrate for economic reasons more frequently than women, while women in-migrate for marriage more often than men. This is consistent with the tradition in rural Vietnam, where women after marriage almost always change residency

Table 11.9 Determinants of under five morbidity

Factor	Level	Adjusted OR [1]	P-Value
Out-migration of father	Yes	1.127	0.065
	No[2]	1	
Out-migration of mother	Yes	1.381	0.000
	No[2]	1	
Household economic level	Non-poor	0.908	0.000
	Poor[2]	1	
Occupation of father	Farmer[2]	1	
	Government staff	0.570	0.000
	Worker	0.629	0.000
	Other	0.414	0.000
Occupation of mother	Farmer[2]	1	
	Government staff	0.627	0.000
	Worker	0.869	0.655
	Other	0.785	0.000

Note: 1. Odds ratios adjusted by controlling for age (as a continuous variable and age**2), sex, ethnic group of the child and education of mother and father. 2. Reference group.

to live with the husband's family. People who have a better economic status or educational level, have a better chance to out- migrate for work in comparison to the poorer people. The care from a mother to her child is very important, which is highlighted the fact that children born to out-migrant mothers have a higher risk of getting sick. Children born to out-migrant fathers do not have a higher risk of getting sick, probably because they have received better care from their mothers and better economic conditions from their fathers. So it is important that there is adequate child care available when a mother out-migrates.

References

Anh, D.N., Tacoli, C. and Thanh, H.X. 2003. *Migration in Vietnam: A Review of Information on Current Trends and Patterns and Their Policy Implications*. Presented at the Regional Conference on Migration, Development and Pro-Poor Policy Choices in Asia.

Cu Chi Loi 2009. *Rural to Urban Migration in Vietnam*, available at http://ideaix03.ide.go.jp/English/Publish/Books/Asedp/pdf/071_cap5.pdf (last accessed 9 April 2009).

Development Analysis Network (DAN) 2001. *Labour Market in Transitional Economies in South East Asia and Thailand.* Development Analysis Network.

Gwatkin, D.R., Rutstein, S., Johnson, K., Pande, R.P. and Wagstaff, A. 2000. *Socio-economic Differences in Health, Nutrition and Population.* Washington: The World Bank.

Vietnam General Statistical Office (GSO) 2004. *Statistical Yearbook 2003*. Hanoi: Statistical Publishing House.

Gwatkin, D.R., Rutstein, S., Johnson, K., Pande, R.P. and Wagstaff, A. 2000. *Socio-economic Differences in Health, Nutrition and Population.* Washington: The World Bank.

Houweling, T.A., Kunst, A.E., Mackenbach, J.P. 2003. Measuring health inequality among children in developing countries: does the choice of economic status matter? *International Journal of Equity in Health*, 2, 8. Available at: http./www.equityhealthj.com/content/2/1/8 (last accessed 9 April 2009).

Winkels, A. and Adger, W.N. 2002. *Sustainable Livelihoods and Migration in Vietnam: The Importance of Social Capital as Access to Resource*. International Symposium on Sustaining Food Security and Managing Natural Resources in Southeast Asia – Challenges for the 21st Century, Chiang Mai, Thailand, 8–11 January 2002.

World Health Organisation 2007. *Country Health Information Profile Revision 2006*. Manila: WHO Western Pacific Regional Office, 373–84.

PART IV
Epilogue

Chapter 12
The Relevance of the Volume

Cheikh Mbacké

HDSS and the Study of Migration

Of all demographic phenomena, migration is the most elusive and difficult to measure. Unlike fertility and mortality which are biologically related to age, migration may occur across a wide range of ages and may be defined with varying criteria that change the number of migrations counted. Censuses and retrospective surveys constitute the major data source for the study of migration, but they share the same limitations as any attempt to capture a dynamic phenomenon using a single snapshot: recall errors including errors in the timing of events and limitations in observing only specific types of movements. It is widely recognized that migration estimates based on census data do not entirely reflect actual, on-going patterns. Health and Demographic Surveillance Systems (HDSS) overcome these limitations through repeated observations at intervals varying from three to twelve months. HDSS are similar to videos that continuously record entries (by birth or in-migration) and exits (by out-migration or death) from the demographic surveillance area (DSA). The preceding chapters demonstrate that HDSS provide a unique opportunity to scrutinize the highly complex social phenomenon that migration is. This chapter summarizes and points to approaches for improving the collection of migration data by HDSS-centred research initiatives.

Assessing Migration Patterns

Migration is highly repeatable and less predictably connected with age than fertility and mortality. However, despite the diversity of contexts (rural and urban, African and Asian) represented in the data and some variation in the threshold of time used to classify migration, there is a relatively regular age structure to migration in these INDEPTH sites. As shown in Chapter 4, the modal group is young adults, sometimes accompanied by children. Labour migration is a key component of these migration profiles, also children accompanying migrant parents and to a lesser extent marriage or marriage dissolution or households moving to access better services. This evidence strongly suggests that the nine-parameter, multi-exponential model of Rogers and Castro (1981: 45) could be a good model to fit the INDEPTH data. Such models can be used to infer migration patterns and to correct observed data by interpolating and smoothing out irregularities. However their extreme complexity makes them less useful in the estimation of migration intensity, stocks or flows.

A thorough and accurate understanding of migration dynamics requires a longitudinal approach. The continuous monitoring of change and the timing of events afford scientists deeper insights into causality because of an increased ability to address migrant selectivity with fewer assumptions, disentangle relationships between interdependent behaviours, choices and events; and deal with lags between the introduction of a source of change and its effects. Being able to do this is critical to the understanding of issues such as the cumulative effects of repeated migration, the impact of changes in exposure (to new ideas, behaviours, practices and diseases) on the migrant and her/his community of origin and the disruptive consequences of separation of family members.

The rapid expansion of HDSS-centred studies since 1990, both in terms of numbers and in population covered (Mbacké and Phillips 2008), attests to their importance as an essential source of health information in the absence of vital registration systems and the chapters in this volume illustrate the contribution of the HDSS approach for getting at the elusive migration element of population dynamics. Each of the preceding chapters focuses on a specific piece of the puzzle and its contributions to unfolding household dynamics in a relatively small area of a developing country in Africa or Asia. Apart from the Nairobi site, all these areas are rural and none can claim to be representative of their host countries. However, as reflected in Chapter 3, the similarities across the socioeconomic systems they represent combined with the near commonality of approach across sites confer a wider scope to the seemingly scattered findings presented in this book. Each gives a locally significant perspective on a phenomenon that is common to all: migration. The findings have important implications both for research and for development policies and programs.

What Have We Learned?

Migration has long been viewed as a major element of household economies, often called the New Economics of Labour Migration theory (Stark and Yitzhaki 1988, Taylor 1999) and in the sustainable livelihoods approach (Scoomes 1998). Chapters 5 to 7 provide empirical support for the new household economics/livelihoods theories by shedding light on the reality of migration as a key livelihoods strategy. They have shown that migration clearly contributes to improved living conditions and social resilience at the place of origin, which is defined as the ability of sending communities to absorb external shocks and stresses without significant upheaval (Adger et al. 2002). In Matlab (Bangladesh Chapter 7) international male migrants ensure better educational outcomes to their children remaining at home than non-migrants. In rural South Africa (Chapter 6) short-term female migrants provide vital support to their families of origin. The South African data show that not only do migrants send money, clothes and food, but that these remittances back to the family left behind are greater (not smaller) among employed female migrants, as compared to their male counterparts. Thus, female migrants are the most vital contributors to the upkeep of the poorest households. The Kanchanaburi chapter (5) unveils rural households' strategies for making most of the tradeoffs between migration and the labour needs of intensified agriculture in rural Thailand. The net effect of sending a migrant away from agriculture on the sending household is not straightforward. It depends on the household's ability to compensate for this loss through labour reallocation and land use adjustments.

The second set of chapters (8–11) provides a different perspective into the intricate links between migration and livelihoods with a focus on health and survival of family members. Three chapters focus on the implications of migration for children under five who represent the most vulnerable fraction of the population and therefore are particularly prone to negative consequences of migration whether they move with or without their parents or are left behind. The Filabavi (Vietnam Chapter 11) data point to the importance of maternal care for these children by revealing a higher incidence of illness among children whose mothers have left them behind. It is important to note that no such negative impact is seen by the out-migration of the children's fathers, underscoring the importance of the mother's role in providing health care to young children. Chapter 8 shows that children who are born in Nairobi's urban slums to non-migrant mothers have significantly higher survival chances than those born to in-migrant mothers, regardless of their origin. On the other hand, the other Kenyan study, conducted at the Nyanza site, shows that migrant children moving from Kenyan urban areas to rural Nyanza enjoy a clear survival advantage compared to both non-migrant and migrant children from other rural areas (Chapter 9). The fact that these findings come from an urban and rural site in the same country epitomizes the complexity of the relationship between migration and child survival. This relationship depends on a host of factors including exposure to new threats, migrant selectivity and differential health endowments between migrants and non-migrants.

The study presented in Chapter 10 portrays the changing dynamics of mortality differentials between returning migrants and non-migrants in Manhiça (Mozambique) and provides insight into the unfolding consequences of the HIV/AIDS epidemic. In general, returning migrants were expected to be positively selected for health and economic situation, retiring with their earnings to the comfort of their home village. HIV/AIDS changed that situation, migrants returning home could be in poor health, needing care and support from their families (Clark et al. 2007). The Manhiça study focuses on this reversal of survival advantages of migrants returning to their home communities and shows that it occurred around 1999, more than a decade after the emergence of HIV/AIDS. This illustrates perfectly the phenomenon of 'migrants returning home to die' that is also documented in neighbouring rural South Africa (Clark et al. 2007). The influx of sick and dying migrants is shifting the burden of disease in affected areas and placing heavy stress on local health care systems.

Thus, the findings in the chapters focusing on migration and health document the potential negative consequences of migration for the health of migrants and overall health of sending communities. These findings contrast with the beneficial impacts of migration unearthed by the chapters dealing with livelihoods. They also exemplify the fact that the impacts of migration can go either way, they can be positive or negative for sending and/or receiving communities depending on the issues at hand and the type of migration under consideration.

Policy Relevance

Development policies need a solid empirical grounding in order to optimize the use of scarce resources. This is particularly important in the poor and marginal settings covered by this book. Given the intensity of population movements recorded in the different sites it is important that policy makers and program designers and implementers understand and take into account migration in their efforts.

Target population Migration changes the composition of the target population in predictable ways. As shown in Chapter 4, young adults are dominant among migrants and policies and programs addressing this critical age group need to recognize their mobility, perhaps being prepared to implement activities in both the origin and destination communities, before and after moving. Better off households are sometimes more likely to reap the benefits of migration and this can contribute to increasing inequalities and this selectivity could be incorporated into poverty-reduction programs by enhancing outreach to those households with no or few migrants. Finally, there might be access issues for in-migrants, particularly those who are disadvantaged and not fully integrated in local households.

Intervention design and implementation The studies in this volume demonstrate the importance of explicitly considering migrants and their families when planning population and health interventions. For example, Chapter 8 clearly points to the need for health programs active in Nairobi's slums to target children whose mother recently arrived in the slum area. Similar attention needs to be paid to Filabavi (Vietnam) children whose mothers have left them behind.

The ability to identify households with out-migrants can help shape interventions in ways that enhance the positive impact of remittances and contributions. Out-migrants who, like the female South African short-term migrants described in Chapter 6, maintain strong links with their families can be effective change agents and interventions should be designed in such a way as to tap this potential whenever possible.

Program evaluation It is clear from the site studies that migration significantly affects our ability to measure program impact. It can lead to heavy attrition of program beneficiaries and introduce serious biases. However, according to

Alderman et al. concerns about the impact of biases due to attrition in longitudinal studies are exaggerated. Their study employed a wide variety of outcome variables from panel surveys in Kenya, South Africa and Bolivia to show that even fairly high attrition rates do not constitute a general and pervasive impediment to attempts to use longitudinal data to control for unobserved fixed factors and to capture dynamic relationships (Alderman 2001: 114).

The studies in this book demonstrate the potential of utilizing the migration data that is necessarily collected with HDSS. A better understanding about how to do this would help all HDSS sites in overcoming a major and persistent criticism that HDSS are under-analysed and underutilized, failing to fully achieve their ultimate goal of providing the evidence base needed to improve public health (Chandramohan et al. 2008). Although incorporating the analysis of migrants into HDSS data analysis strategies is only one of many factors affecting HDSS analytic productivity, it is hoped that the contributions in this volume illustrate the vast possibilities that would arise from greater HDSS data sharing and utilization. As demonstrated in this volume, developing country scientists with an intimate understanding of local population movements have enhanced HDSS contribution to improved development policy and practice in developing countries.

Improving the Collection of Migration Data at INDEPTH Sites

Most HDSS were developed to study the impact of health interventions but their comprehensive data collection system extends their usefulness far beyond health. The continuous recording of moves in and out of the DSA provides the foundation for a unique contribution to unravelling some aspects of population movements that have remained obscure until now because of lack of adequate data.

In order to improve their ability to study migration and to reap the full collective potential of HDSS sites, some improvements are needed in the collection of migration data by INDEPTH member sites.

First of all, sites need to follow a simple procedure at all scheduled household visits to record:

- Arrival date, origin (place of birth and/or last residence), reason and intention to stay for all people who spent the preceding night in the house but were not present at the preceding visit.
- Departure date, destination, reason and intention to return for those who left the household since the last visit.

These records must be reconciled regularly in order to identify multiple movements by the same individual. Once this basic procedure is respected, the fundamental raw material will be available and the definition of migration (however complex) can be done a posteriori (Picouet 1971) and standardized as needed across sites. Provided that each individual has a unique identification number, a single cross-sectional survey using the same instrument in multiple sites can produce groundbreaking answers to some of the questions that still remain intractable to migration experts. The fact that data from any study fielded in a HDSS site can be linked to individuals under observation and hence build on the foundation provided by previous studies permits the accumulation of scientific capabilities and serendipitous research (Mbacké and Phillips 2008).

Finally, we should always keep in mind that HDSS are relatively expensive and that current efforts to reduce their cost without jeopardizing data quality should be strengthened. If successful, the INDEPTH Network's drive for the adoption of new technology to improve data collection, management and sharing at member sites will go a long way in reducing HDSS per capita operating costs and freeing more resources for service delivery.

What Have We Missed and Why?

Rather than treating migration as a by-product of the need to have accurate population counts at each surveillance site, the chapters in this volume demonstrate the potential of explicitly focusing on migrants. But very little is known about the in-migrants prior to their arrival in the study areas, nor of what happens to the out-migrants who have left the area. The study reporting on temporary or circular migration (South Africa) hints at the richness that can be obtained with a more complete tracking of people in and out of the area. In order to fully exploit this foundation, routine HDSS data must be complemented by specialized surveys that capture aspects that are difficult to explore during regular household visits. Such surveys could provide an in-depth look at critical issues such as the nature of activities at destination, distance of migration and frequency of links with sending families, impact on migration of other family members and remittances. Given the concern for many INDEPTH member sites to monitor changes in health behaviour, such surveys could be used to measure changes in health behaviour, including those subject to intervention in the DSA. Because HDSS do not attempt to follow migrants in their destination, they will always miss some of the effects of migration that can only be captured in the place of destination.

Follow-up surveys could be conducted in the destinations most common for out-migrants from the DSA to cover issues of changes in health status, behaviour, as well as economic behaviours. With these carefully targeted surveys in the DSA and destination areas INDEPTH researchers would be well-positioned to address some of the critical issues raised in this volume regarding migration, health and livelihoods.

References

Adger, W.N., Kelly, P.M., Winkels, A., Huy, L.Q. and Locke, C. 2002. Migration, remittances, livelihood trajectories and social resilience. *Ambio*, 31(4), 358–66.

Alderman, H., Behrman, J.R., Kohler, H.P., Maluccio, J.A. and Watkins, S.C. 2001. Attrition in longitudinal household survey data. *Demographic Research*, 5(4), 79–124.

Chandramohan, D., Shibuya, K., Setel, P., Cairncross, S., Lopez, A.D. et al. 2008. Should data from demographic surveillance systems be made more widely available to researchers? *Public Library of Science (PLoS) Medicine* 5(2): e57. doi:10.1371/journal.pmed.0050057.

Clark, S.J., Collinson, M., Kahn, K., Drullinger, K. and Tollman, S. 2007. Returning home to die: circular labour migration and mortality in South Africa. *Scandinavian Journal of Public Health*, 35(Suppl. 69), 35–44.

Mbacké, C.S.M. and Phillips, J.P. 2008. Longitudinal community studies in Africa: challenges and contributions to health research. *Asia-Pacific Population Journal*, 23(3), 23–38.

Picouet, M. 1971. Observation des migrations intérieures dans les pays a statistiques incomplètes. *Cahiers de l'ORSTOM, Série Sciences Humaines*, 8(1), 37–49.

Rogers, A. and Castro, L.J. 1981. *Model Migration Schedules* – Research Report – 81–30. Laxenburg, Austria: International Institute for Applied Systems Analysis.

Scoomes, I. 1998. *Sustainable Rural Livelihoods: A Framework for Analysis*. IDS Working Paper. I.F.D. Studies. Brighton: University of Sussex.

Stark, O. and Yitzhaki, S. 1988. Labour migration as a response to relative deprivation. *Journal of Population Economics*, 1(1), 57–70.

Taylor, J. 1999. The new economics of labour migration and the role of remittances in the migration process. *International Migration*, 37(1), 63–86.

Index

adult mortality in Mozambique 159–68
African Population and Health Research Centre (APHRC) 125
Age Specific Fertility Rate (ASER) 25
age–sex migration profiles
  conclusions 61
  female 55–7
  INDEPTH sites 49, 53–7
  introduction 51–3
  Kenya (Kisumu) 51–3
  levels comparison 59–61
  male 53–5
  profile 51–3
  temporary migration 57–59
agriculture in Kanchanaburi, Thailand 71–5, 75–8
APHRC *see* African Population and Health Research Centre
ASER *see* Age Specific Fertility Rate
attrition model for childhood mortality in Nairobi 129–32

Bangladesh (Matlab)
  parents' migration/children's education
    background 109–10
    conclusions 117
    discussion 116–17
    HDSS 110–11
    methods/data 110–13
    results 113–16
  book summary 9–10, 13–15

children
  education in Matlab, Bangladesh
    background 109–10
    conclusions 117
    discussion 116–7
    methods/data 110–13
    results 113–16
migration and mortality in Kisumu, Kenya
  discussion/conclusions 150–4
  introduction 139–45
  malnutrition 151
  methods 145–7
  results 148–150
  water supply 150, 151
migration and under five mortality in Bavi, Vietnam
  conclusions 180–1
  introduction 169–70
  methodology 170–3
  results/discussion 173–81
mother's migration and mortality in Nairobi, Kenya
  conclusions 135–6
  discussion 133–5
  introduction 123–5
  limitations 135
  results 129–33
  study population/methods 125–9
CHS *see* community health stations
communication and migration 8–9
community context
  health 34–6, 44–6
  livelihoods 44–6
  migration in INDEPTH sites 34–6, 36–44
community health stations (CHS) 171
cross-site comparative analyses and HDSS 30–1

Demographic and Health Survey (DHS) 124
demographic surveillance 3–18
demographic surveillance area (DSA)
  HDSS 5–6, 16, 28, 185
  health status 190
  household definition 31
  Kenya (Kisumu) 139–40, 144–7, 148–53

destinations and migration 7–8
DH *see* district hospital
DHS *see* Demographic and Health Survey
disease exposure and migration 11–12
district hospital (DH) 171
Doi Moi (liberal policies in Vietnam) 169
DSA *see* demographic surveillance area
economics and migration 66–9

education
  children in Bangladesh 109–17
  mortality 166
Expanded Programme on Immunization (EPI) in Kenya 139

HDSS *see* Health and Demographic Surveillance Systems
health
  community context 34–6, 36–44, 44–6
  demographic surveillance migration 19–32
    concepts 21–2
    HDSS 22–3
    measurement 20–2
    migration 10–13, 13–15
Health and Demographic Surveillance Systems (HDSS)
  age profiles 15,
  age–sex profiles 54–7, 59–61
  Bangladesh 110–11
  census 19–20
  characteristics 3–5
  cross-site comparison 30–1
  data 26–9
  data collection 5, 189–90
  developing countries 6–8
  Kenya
    Kisumu 13, 51–3, 145, 150, 152
    Nairobi 125–9
  methods 23–6, 29–30
  migration
    accuracy 185
    data errors 28–9
    data/methods 23–7
    health 10–13, 13–15
    importance 186
    labour 58
    livelihoods 6
    remittances 6–9
    surveillance 4–5
    tracking 5, 15–16, 57–8
  Mozambique (Manhiça) 159–60, 161–6
  omissions 190
  outlines 22–3
  programme evaluation 188–9
  sites 9–11, 13–15, 16, 21–2
  socio-economic status 8
  South Africa 86, 88
  Thailand (Kanchanaburi) 69
    HIV/AIDS 49, 89, 159, 166, 187
    households in Kanchanaburi, Thailand 69–78

IMCI see Integrated Management of Childhood Illness
immunization in Kenya 139
INDEPTH networks
  age–sex migration profiles 49–51, 53–7, 59–62
  collection of data 189–90
  community context of migration 34–6, 36–44, 45–6
  Data System Initiative 31
  fertility study 49
  migration patterns 185–6
  omissions 190
  sites 3, 19, 23, 29, 30
information and migration 8–9
in-migration (HDSS sites) 52, 55, 57
Institute for Population and Social Research (IPSR), Thailand 69
Integrated Management of Childhood Illness (IMCI), Kenya 139

Kenya
  Expanded Programme on Immunization 139
  Integrated Management of Childhood Illness 139
  Primary Health Care 139
Kenya (Kisumu, Nyanza Province)
  child migration and mortality
    conclusions 150–4
    HDSS 145, 150
    introduction 139–45

methods 145–7
results 148–50
water supply 150, 151
migration 51–3
KwaZulu–Natal Income Dynamics Study (KIDS) 84

labour and migration 58, 165
land use and agriculture 74–7
LTFU *see* loss-to follow-up
liberal policies (Doi Moi) in Vietnam 169
livelihoods and community context 44–6
longitudinal analysis of data 152
loss-to-follow-up (LTFU) of departures 4–5, 26, 28, 146

malnutrition in children in Kenya 151
Manhiça Health Research Centre, Mozambique 160
men and age profiles in HDSS sites 53–5
migration
adult mortality (Manhiça, Mozambique) 159–68
age profiles 49–51, 51–62
agriculture (Kanchanaburi, Thailand) 65–80
child mortality (Kenya) 140–5, 145–50, 148–54
communication 8–9
community context 34–6, 36–44
destinations 7–8
disease exposure 11–12
economics 66–9
HDSS 4–9, 15–16, 23–9, 185–6
health 10–13, 13–15
INDEPTH networks 185
information 8–9
Kenya (Kisumu) 51–3
labour 58, 165
levels comparison 59–62
livelihoods 6–9, 9–10
mother's migration/childhood mortality (Nairobi, Kenya) 123–37
outcomes 186–9
parents' migration/children's education (Matlab, Bangladesh) 109–19
permanent 57–8
remittances 6–7, 7–9
sex profiles 49–51, 51–62
socio-economic change (South Africa) 81–108
socio-economic status (South Africa) 81–3, 95–100, 101–5
stress 11
surveillance 3–18
temporary 57–9
under five mortality in Bavi, Vietnam 169–81
vigour 12–13
Migration and Urbanization Working Group (MUWG) 30
mother's migration and childhood mortality in Nairobi 123–37
Mozambique (Manhiça)
migration and adult mortality
background 159–60
conclusions 166–7
HDSS 159–60, 161–6
methods 160–1
results 161–5
MUWG *see* Migration and Urbanization Working Group

Nairobi, Kenya
mother's migration and childhood mortality
conclusions 135–6
discussion 139–5
HDSS 125–9
introduction 123–5
limitations 135
population/methods 125–9
results 129–33
study population/methods 125–9
Nairobi Urban Health and Demographic Surveillance System (NUHDSS) 125
NELM *see* New Economics of Labour Migration
New Economics of Labour Migration (NELM) 6, 81
NUHDSS *see* Nairobi Urban Health and Demographic Surveillance System

out-migration
HDSS sites 52, 54, 56

Thailand (Kanchanaburi) 73–4
Vietnam (Bavi) 181

parents' migration in Matlab, Bangladesh 109–19
PCA *see* Principal Component Analysis
permanent migration 57–8
PHC *see* Primary Health Care
poverty in South Africa 83–5
Primary Health Care (PHC) in Kenya 139
Principal Component Analysis (PCA) 172–3
Project for Statistics and Living Standards and Development (PSLSD) 83

remittances and migration 6–7, 7–9

sex profiles and migration 49–51, 51–62
socio-economic status (SES) in South Africa 81–3, 85, 87–90, 95–105
South Africa migration
2000–2007 81–108
data/methods 86–9
HDSS 86, 88
introduction 81–5
poverty 83–5
remittances 2002–2007 91–4
results 89–101
socio-economic changes 2001–2005 89–90
socio-economic status 81–3, 85, 87–90, 95–105
stress and migration 11
surveillance, demographic 3–18
see also demographic surveillance area

Taylor, J.E. 6
temporary migration 57–9
Thailand, (Kanchanaburi)
agriculture 75–8
conclusions 78
data/methods 69–71
HDSS 69
household survey 69–78
introduction 65–9
migration 65–80
results/discussion 71–8

Vietnam
demographics 169
liberal policies (Doi Moi) 169
Vietnam (Bavi)
migration and under five mortality
conclusions 180–1
introduction 169–70
methods 170–3
results/discussion 173–80
vigour and migration 12–13

water supply in Kenya 150, 151
women
age profiles 55–7
agriculture 75